# How Much E[illegible]
# *Your* Building Use?

Liz Reason with Kerry Mashford

Email: **liz@lizreason.co.uk**

Telephone: 01608 811212

Mobile: 0771 340 7772

First published in 2014 by Dō Sustainability
87 Lonsdale Road, Oxford OX2 7ET, UK



ISBN 978-1-910174-04-3 (eBook-ePub)
ISBN 978-1-910174-05-0 (eBook-PDF)
ISBN 978-1-910174-03-6 (Paperback)

A catalogue record for this title is available from the British Library.

Dō Sustainability strives for net positive social and environmental impact. See our sustainability policy at **www.dosustainability.com.**

Page design and typesetting by Alison Rayner
Cover by Becky Chilcott

For further information on Dō Sustainability, visit our website:
**www.dosustainability.com**

# DōShorts

**Dō Sustainability** is the publisher of **DōShorts**: short, high-value ebooks that distil sustainability best practice and business insights for busy, results-driven professionals. Each DōShort can be read in 90 minutes.

## New and forthcoming DōShorts – stay up to date

We publish 3 to 5 new DōShorts each month. The best way to keep up to date? Sign up to our short, monthly newsletter. Go to **www.dosustainability.com/newsletter** to sign up to the Dō Newsletter. Some of our latest and forthcoming titles include:

- ***Strategic Sustainability: Why it Matters to Your Business and How to Make it Happen*** Alexandra McKay
- ***Sustainability Decoded: How to Unlock Profit Through the Value Chain*** Laura Musikanski
- ***Working Collaboratively: A Practical Guide to Achieving More*** Penny Walker
- ***Understanding G4: The Concise Guide to Next Generation Sustainability Reporting*** Elaine Cohen
- ***Leading Sustainable Innovation*** Nick Coad & Paul Pritchard
- ***Leadership for Sustainability and Change*** Cynthia Scott & Tammy Esteves
- ***The Social Licence to Operate: Your Management Framework for Complex Times*** Leeora Black
- ***Building a Sustainable Supply Chain*** Gareth Kane
- ***Management Systems for Sustainability: How to Successfully Connect Strategy and Action*** Phil Cumming

- ***Understanding Integrated Reporting: The Concise Guide to Integrated Thinking and the Future of Corporate Reporting*** Carol Adams
- ***Corporate Sustainability in India: A Practical Guide for Multinationals*** Caroline Twigg
- ***Networks for Sustainability: Harnessing People Power to Deliver Your Goals*** Sarah Holloway
- ***Making Sustainability Matter: How To Make Materiality Drive Profit, Strategy and Communications*** Dwayne Baraka
- ***Creating a Sustainable Brand: A Guide to Growing the Sustainability Top Line*** Henk Campher
- ***Cultivating System Change: A Practitioner's Companion*** Anna Birney

## Subscriptions

In addition to individual sales of our ebooks, we now offer subscriptions. Access 60+ ebooks for the price of 5 with a personal subscription to our full e-library. Institutional subscriptions are also available for your staff or students. Visit **www.dosustainability.com/books/subscriptions** or email **veruschka@dosustainability.com**

## Write for us, or suggest a DōShort

Please visit **www.dosustainability.com** for our full publishing programme. If you don't find what you need, write for us! Or suggest a DōShort on our website. We look forward to hearing from you.

# Abstract

**WHY DO AWARD-WINNING 'GREEN' BUILDINGS** so often have higher energy bills than ordinary buildings? Why do expensive refurbishments deliver outcomes that are far from the promises of improved sustainability? Why does your building have high running costs and still the occupants complain about being too cold or too hot and are otherwise dissatisfied?

The UK's failure to produce buildings that are comfortable with excellent energy performance should be regarded as a national scandal. Yet achieving low energy buildings does not involve learning rocket science. Just some basic building physics, a clear language for talking meaningfully about energy-efficient outcomes with all those in the buildings cycle, and an outlook that casts a new low energy perspective on old problems.

This book provides that common language. It outlines a path towards understanding what makes for a good quality low energy building, the stakeholders that need to be engaged, and encourages new ways of thinking about how to reduce energy use and costs.

# About the Author

**LIZ REASON** is Managing Director of the **Green Gauge Trust**, a not-for-profit which aims to mainstream the knowledge and skills for low energy low carbon buildings. Its Green Stripes programme encapsulates the core curriculum for understanding the key principles and practices for delivering low energy low carbon buildings.

Liz has been a Director of the National Energy Foundation since 2006 and a trustee of the charity, SuperHomes. In April 2012, Liz was appointed to the Knowledge & Skills Working Group of the BIS Green Construction Board.

She was on the Development Panel of Octavia Housing from 2010-2013. From 2004-2009, Liz was director of the AECB's CarbonLite. In 2009-2010 she helped to set up what is now the Passivhaus Trust.

She has been both an adviser and an assessor for a number of Technology Strategy Board buildings competitions. In 2008-2009, she was a member of the government's Eco-towns Challenge Panel.

www.lizreason.co.uk

---

**DR KERRY MASHFORD**, CEng FIMechE FRSA has been Chief Executive of the **National Energy Foundation** since December 2012. Her previous roles include Lead Technologist in low impact buildings at the Technology

Strategy Board, Technical Director of the Ecological Sequestration Trust, Director of Development for the Centre for Remanufacturing and Reuse, Director of Sustainable Development at Benoy Architects and Masterplanners, and Head of Sustainable Manufacturing & Construction for ARUP. Kerry also co-founded Interfacing Ltd., an engineering design and consultancy company focusing on sustainable projects, materials, technologies and systems.

Kerry has an MSc in interdisciplinary design of the built environment (IDBE) from the University of Cambridge and a PhD in integrated design of complex, multi-domain systems from Brunel University. A Chartered Engineer, Fellow of the Institution of Mechanical Engineers, and Fellow of the Royal Society of Arts, Kerry is an advisor to several academic institutions, a member of IMechE Council and a lead assessor for the Manufacturing Excellence Awards and judge of several energy and buildings related awards.

# Acknowledgments

**LIZ REASON THANKS** Bill Bordass for all that he has taught her, much of it captured in this book. No doubt there will be things which do not meet Bill's own high standards, but she hopes he will forgive her in the interests of reaching a wider readership!

# Glossary

**Absolute energy target:** one expressed as a specific number e.g. 15 kWh/m$^2$/yr

**EPBD:** Energy Performance of Buildings Directive

**EPC:** Energy Performance Certificate

**DEC:** Display Energy Certificate

**ESOS:** Energy Saving Opportunities Scheme

**Performance gap:** the difference between design and actual energy use

**Relative energy target:** one expressed as a percentage reduction against another number (which has usually not been measured)

# Preface

**IT WAS WAY BACK IN 1987**, when I had just stopped full-time teaching, that John Walker – then Planning Director at Milton Keynes Development Corporation – approached me, together with a colleague, Steve Fuller, to ask if I would chair a new charity he wanted to create. Its mission would be to disseminate widely across the UK the expertise the Corporation had acquired in commissioning a then-unique cluster of energy-efficient dwellings in the new town. I was pleased to be asked because my research interests in solar energy conversion had given me a lively appreciation of the importance of attending to energy efficiency before adding in any renewables. Thus the National Energy Foundation was born, and I became its first chairman, and Steve its first CEO.

The rest, regrettably, is not history. John is now chairman of the NEF, and I am its president; but the Foundation's message that buildings that waste energy are costly to run, uncomfortable to live in and add an avoidable burden to climate change is as much needed as ever. We all know that old buildings were not built with energy efficiency in mind and will usually need retrofitting to reach an acceptable standard, but what is not widely understood is that even modern buildings designed to energy-efficient specifications often do not perform as they should.

Hence this book, co-written by NEF trustee Liz Reason with Chief Executive Kerry Mashford. Although it may seem extraordinary that it should need to be written, there will be few people or organisations that can answer the question posed by its title. Read on, and you will understand why!

Liz has set out clearly what has stopped us from delivering high quality, energy-efficient buildings to date and what we need to do to change that; Kerry has added her wealth of professional experience, most recently with Technology Strategy Board, where her focus was primarily on building performance evaluation.

For me, this book epitomises what the National Energy Foundation is about – helping people take control of the energy use in their buildings. Enjoy it!

***Mary Archer***
February 2014

# Contents

# Foreword

**I THOUGHT YOU SHOULD KNOW** that I am a modern linguist. I'm telling you this because too many people respond to discussions about energy and buildings by saying "I'm not technical" (sub-text: "This will be far beyond my ability to understand.")

We should not confuse 'technical' with 'practical'. Believe me, I'm not very practical. I can barely bang a nail into a piece of wood. I'm a huge admirer of those who make things, and particularly those who design and construct buildings. It's not so much the feats of engineering, but knowing what to do when. All those critical paths where one thing can't happen before the other – but the other doesn't turn up on site! But understanding the technical aspects of buildings and what constitutes excellent energy performance does not mean that you have to be able to deliver it.

I went to a school that prided itself on educating the 'crème de la crème'. No doubt relying on the notion of cream always rising to the top, they gave no career guidance, other than the expectation that you should go to university, whether you had a clear idea of what you wanted to do or not. As a result, like many people, I left university with a degree I had found boring and no idea about how I might spend the rest of my life. So I did two years' voluntary service overseas. Then I trained to be a teacher. And then I met Daisy the Cow.

Daisy was a cow of two halves – a man in the front legs and a woman in the back. She was walking the streets of Cambridge to make a plea for vegetarianism. This was not the “I’m a cuddly animal, please don’t eat me” kind of a plea. Daisy’s rationale was that she ate a lot of grain and was an energy-intensive way of supplying humans with food – 7lbs of grain in fact, to make 1lb of edible Daisy.

I have to say that this was news to me. And as someone who’d never really stopped to think about whether I should eat meat or not, it’s a message that packed a punch. I’d lived in Africa, so I knew that we in the developed countries took more than our fair share of resources and, what’s more, took it for granted that we should. So I stopped to find out about who’d brought Daisy to the streets, and found myself talking to Cambridge Friends of the Earth (FoE).

The key message that FoE was out to promote was that we use all our resources without reflecting on the impact on other humans and on the planet. This made sense to me. So I decided to attend one of FoE’s meetings to find out more. And that’s when I discovered energy policy.

Now if there’s one thing those living in the twentieth century took for granted, it was energy. Ownership of cars was designed to allow all of us to travel further, faster, and more frequently. Our homes would be warm all the time, requiring us only to turn up the thermostat, not carry in the coal. We could live in a world of constant light and on-stream entertainment. Consuming fossil fuels like there was no tomorrow, consuming calories like food producers were going out of business. By the beginning of the twenty-first century, air travel had become ‘a human right’; in Summer we put on jumpers and jacket to go into the office from outside, because “no company will rent a building without air-conditioning”.

At the time I joined FoE, it was embroiled in the Windscale Inquiry, putting a professional and rationally argued case against the development of the site in Cumbria for more re-processing of nuclear waste. Fundamental to its arguments was one of cost. When the first nuclear power station was opened, the headlines read "Nuclear power – too cheap to meter!" FoE argued that this would not be the case, that nuclear power was a complex technology, fraught with environmental risks, and subject to unsubstantiated cost claims from the state-owned Central Electricity Generating Board. How much cheaper it would be simply to use less energy – the term 'negawatts' was later coined to describe the low-cost reduction in megawatts of electricity demand.

To me this all proved a heady brew and I was hooked. Energy – we use it for everything, we take it for granted, and it will probably be our nemesis.

So for the next few years, I set about re-defining myself from a modern linguist, to an energy policy analyst, with jobs in fuel poverty, energy efficiency, consumer representation, wind generation, energy markets, and the built environment.

These all helped me to understand different aspects of the policy process, including the ways in which we create problems, which then have to be solved. For example, developing a carbon calculator for school designers that allowed them to claim that a building was zero carbon simply by installing a biomass boiler was clearly a gross simplification that put headline targets ahead of basic science and practicalities. As a result, head teachers and school building managers have become the unhappy operators of sometimes complex and tricky heating systems, with biomass boilers running in parallel with gas heating systems.

Or, in many leases, building owners pay the energy bills and tenants have little incentive to save energy in their leased space. Where energy costs are paid directly by tenants, building owners aren't driven to invest in efficient building systems. This dynamic is commonly referred to as the "split incentive" barrier to energy efficiency.

*Throwing ill-thought out solutions to problems is no answer. That's not a matter of being technical – that's just common sense. We have to plan, design, construct and manage buildings carefully if we are to deliver improved energy and carbon performance. This book aims to show you how.*

# Introduction

**YOU MAY BE ASKING YOURSELF** "Why would I know how much energy my building uses? Nobody's ever asked me that question before. Why do I need to know?"

Imagine the question had been asked about your car. Just search on "mpg" and choose amongst the multiple websites helping you establish how much fuel every make of car uses, how they compare with one another, how your mpg compares to that of other real users who drive the same car as you, in the same area. Car advertisements must include measured energy use by law (albeit with the current methodology subject to scrutiny and change to make the figures more realistic).

It's not so surprising, is it? In fact, we take this for granted.

This book is for a business audience, but we all live in homes. The average price paid for a car these days is around £15,000. The average price of a new house (March 2014) is £172,000 – twelve times higher. The average mortgage repayment is £494, the average monthly domestic energy bill is £116, or more than a fifth of the total mortgage repayment. But how many people know the mpg of their building? Many people are worried about rising energy bills, yet do not know how little they would need to spend if they made their homes energy-efficient – and what's more, they would not be able to find the answer to that question readily.

When it comes to commercial and public buildings, hundreds of thousands or millions of pounds are invested. You would expect therefore, that those

**http://www.fuelmileage.co.uk/**

**What is fuel consumption?** – Fuel consumption is the amount of fuel a car's engine uses measured in either miles per gallon or litres per 100km.

Compare mpg fuel consumption and average mpg ratings by browsing our list of car makes (**www.fuelmileage.co.uk/list-manufacturers**). Research gas mileage with metric L/100km and imperial mpg as well as car tax bands and car $CO_2$ emissions details.

You can compare fuel economy and average mpg for new cars or use the car mileage search (**www.fuelmileage.co.uk/advanced-search**) to find specific car performance specs. To work out your actual real world fuel economy try using the fuel economy calculator (**www.fuelmileage.co.uk/fuel-mileage-calculator**). For a quick search by car make or car model try one of the search options below.

**http://thempg.co.uk/**

Before buying your car find out it's real mpg from drivers in your town!

**HONDA Accord**

Real combined mpg – 40.4 mpg urban

investing would take a keen interest in the quality of the product they are buying. Will its occupants benefit from comfortable temperatures in both winter and summer? Will the air quality be good? Is noise transmission a problem? Is there plenty of natural daylight? Is the space flexible? Do staff like working in this building?

It's more likely that the building will be sold with a number of features. Like air-conditioning. An atrium. Twelve 24-passenger glass lifts. But what does this actually mean? What are these features supposed to be telling you about the property?

Air-conditioning is probably shorthand for "this is a modern building". What it often means is that people have to wear more clothing layers in summer than in winter because the internal temperature is set so low. Or that the heating and air-conditioning systems are not operated properly and compete with one another so that none of the occupants feels comfortable.

> ***"We moved into an office that was ten years old and had modern features like air-conditioning – and lots of west-facing glass. The overheating in summer was sometimes unbearable. It was only after our prolonged pestering of the building agents that they eventually established that the air-conditioning system had never been properly commissioned and all the tenants had been paying for an ineffective and inefficient system since the building was opened."***

Atrium means 'prestige', expensive, something a successful company can afford. Or, as one manager of a new academy reported: "At the last minute, the cost consultant substituted sensor-operated sliding doors for the revolving doors – at both ends of the atrium – so the main lobby is a wind tunnel."

In an era when corporate responsibility increasingly focuses on sustainability, 'green' attributes might be expected to feature more prominently. Bicycle racks and showers are referenced quite frequently. BREEAM ratings are occasionally mentioned. But the real, measurable performance ratings, such as temperatures achieved and air quality standards in summer and winter, or occupant satisfaction, simply do not feature. Let alone how much energy the building uses to meet a given comfort level, and what are the resulting emissions.

However, since April 2012 it has been mandatory to provide a building's Energy Performance Certificate (EPC) to prospective buyers or tenants when they are viewing a property or when a request to view has been made. (An EPC provides an asset rating for a building – or its design energy performance which, incidentally, may not be a true reflection of its actual energy performance – of which more later). A quick review of property letting and sales websites reveals that, in spite of this requirement, EPCs are not used for marketing or selecting buildings – the providers don't shout about it and customers don't ask for it.

Display Energy Certificates (DEC) (reflecting actual energy use in the building) and an accompanying advisory report, are also required for buildings over 500m$^2$ that are occupied in whole or part by public authorities and which are frequently visited by the public, and over 1000m$^2$ if publicly-owned and frequently visited by the public. So, for example, a school must comply with this legislation, but a hotel need not, although commercial buildings may display a certificate on a voluntary basis[1]. Based on actual energy use, DECs provide an operational rating which has to be displayed in a prominent place clearly visible to the public.

Research published in 2013 showed an interesting relationship between EPC ratings and occupant satisfaction. Although the authors say that they have reservations about the benefit of EPCs as a proxy for wider environmental performance, they conclude that: "EPCs were found to provide valid indicators of expected holistic occupier satisfaction for occupiers, agents and building owners of facilities."[2]. In the absence of other publicly available measures of occupant satisfaction, this could make sense, if we consider that energy is just a vector for comfort and services. A well-managed building is likely to be comfortable and have lower energy bills. So a high-rated EPC (or DEC) could demonstrate that you are getting best value-for-money, have lower costs, higher productivity, and enhanced corporate responsibility.

That's why we are asking how much energy your building uses. Because knowing the answer to that question indicates a lot more than might seem likely at face value. It means that you can benefit from lower costs, higher comfort and productivity, business continuity by reducing the risks of disruption from energy supply shortages, and, for larger organisations, carbon compliance. Lowering your energy use and carbon emissions also means that you can improve corporate and social responsibility performance. And, if you can answer the question, you will have developed knowledge and skills that are relevant to the new energy and climate change future.

So when we talk about 'your' building, which 'you' do we have in mind? You might simply be a building occupant, a client about to invest in a building (one to build, refurbish or rent), a member of the design team involved in delivering a new or refurbished building, a contractor or a building manager getting to grips with energy costs and comfort.

The overall aim of this book is to contribute to improving the energy- and carbon-literacy of those in the property and construction sectors. Given their size and reach – who isn't a building occupant? – that means 'everyone'!

The objectives of this book are threefold:

i. to give you a different perspective on your buildings and how focusing on energy actually gets you a better quality building all round;

ii. to improve your understanding of energy and buildings – equipping you to question building energy use and draw informed conclusions from the answers you receive;

iii. to help you to communicate more effectively about energy use in your building;

iv. to help all readers understand how they can play a part in achieving a low energy building – whether you are:

- investing in new or existing buildings;
- marketing new or existing buildings;
- part of a design and construction team;
- an asset or building manager, managing agent, or building occupant.

The book presents a common language to make possible meaningful discussion about building performance. It explains the gap between design intent and out-turn performance, and why it exists. It helps close that gap by ensuring that clients know what to ask for, and how to establish whether buildings will or do perform as promised. It will help members of design and construction teams and their supply chains alter their perspective from one of delivering a building to a given design standard, to one of delivering to a given performance outcome. And it demonstrates how feedback from users helps inform building management and use.

**Part 1** clarifies the issue of language.

**Part 2** describes ‘the performance gap’.

**Part 3** explains how the structure and culture of the UK property and construction sectors conspire to divert attention from the in-use energy performance of buildings.

**Part 4** demonstrates what you can do to get better buildings.

**Part 5** makes the case for doing more of it and inspires us all by providing examples of what others have achieved whatever their roles in the building design, delivery and management spectrum.

PART 1

# Learning a Common Language

## Energy and carbon are often confused

**TALKING ABOUT LOW ENERGY BUILDINGS**, the director of a major UK multidisciplinary built environment company said: *"Oh yes, we've been involved in a number of low energy buildings. Did a big one in Islington last year...* (puzzled frown)*...uses a lot of energy though".*

This kind of confusion is not unusual. Energy and carbon are believed to be synonymous but they are not. And a low carbon building is most definitely not necessarily low energy.

In this case, it has probably arisen for one of two reasons. Either the building has been designed as 'low *energy*', but is not performing as expected, or it has been designed to be 'low *carbon*' and he is confused about the difference between energy and carbon.

This director is to be congratulated on one thing – the company has clearly set itself a target, and has also been measuring the actual against the intended performance.

Unfortunately, the terms energy and carbon are frequently used interchangeably, when they are two very different things. Understanding the distinction between them is very important if the right decisions are to be made about investments and creating the business case to reduce energy demand, improve energy efficiency, or install renewable technologies.

Why is the distinction not clear? After all, carbon dioxide is a gas that is emitted when fossil fuels are burned. Energy is just – energy – isn't it?

Let's take the electric car as an example of how loose talk misleads. Electric cars are referred to as "emission free" (e.g. Daily Mail 5/9/2013). In reality, they lead to a significant reduction in local air pollution (nitrous and sulphur dioxides), as they do not emit tailpipe pollutants unlike conventional internal combustion engines. However, if they are charged using grid electricity, which they almost invariably are, then they are not contributing a significant saving in greenhouse gas emissions. Grid electricity generation is mainly fossil-fuelled and only 35% efficient overall, using three times as much *primary* energy as it produces in the form of end-use electricity[3]. So electric cars are not responsible for carbon emissions on the street, but they are responsible for carbon emissions at the national grid level (unless they are recharged using zero carbon technologies).

In the same way, heat pumps are often referred to as providing 'free' energy from the ground or air. However, if you want to use a heat pump to reduce carbon emissions, it will need to generate almost three times as much useful space heating energy as the electrical energy used to drive the system (including auxiliary equipment such as pumps and immersion heaters) if they are using grid electricity. (This is the Coefficient of Performance or CoP). Many installed systems do not currently achieve this.

The Energy Saving Trust has undertaken two studies to measure the performance of a large number of air and ground source heat pumps in field trials[4]. The first, published in 2010, analysed the results from 83 installations. Few of the heat pumps were performing as expected, with problems identified as being with one or more of specification, design, installation, commissioning or operation. A subsequent study, published in July 2013, reported the results of a series of major and minor interventions made to optimise the performance of the same heat pumps to deliver the best possible outcome[5]. The average CoP for ground source heat pumps was found to be 2.82 and for an air source heat pump was found to be 2.45. Only a few of the heat pumps exceeded a CoP of 3. Because heat pumps deliver low temperature heat - more bluntly, lukewarm water - it is important that they are installed in buildings with a high quality fabric so the heat doesn't just leak away. They also need large heat emitters - underfloor heating or enlarged radiators - to work at a high CoP.

## The treatment of energy and carbon in the regulatory framework

How do we reflect this distinction between energy and carbon in our regulatory framework? In a rather confused way...

Table 1 sets out the different standards, both statutory and voluntary, which apply to new and refurbished buildings.

The table partly serves to elucidate, but also illustrates that the many different ways of talking about the energy and carbon performance of buildings are a feature of the regulatory framework. And that they use terminology that confuses rather than clarifies. So if some of the following does not seem instantly meaningful, that's because it is unnecessarily complex. These include $CO_2$ emissions as the headline indicator, EPC ratings with A–G and numerical scales, and display *energy* certificates where the main metric illustrated in the A–G grades is actually $CO_2$ rather than energy. It is not a given that a building with low carbon emissions is low energy, if higher energy demand is being met by zero carbon technologies, simply substituting higher emissions energy with lower emissions energy.

**TABLE 1: Key elements of the regulatory framework**

| Name of standard, voluntary or mandatory | Form of rating or compliance target | Measure |
|---|---|---|
| *Building Regulations 2010* | Building Emissions Rate (BER) expressed as a % reduction against the Target Emissions Rate whose calculation is based on a notional building of the same size and shape as the actual building but with properties set out in the 2010 NCM Modelling Guide (Simplified Building Energy Model or SBEM) | $CO_2/m^2/yr$ |
| | % reduction | $CO_2$ in $kg/m^2/yr$ |

| Name of standard, voluntary or mandatory | **Form of rating or compliance target** | **Measure** |
|---|---|---|
| ***Building Regulations 2013*** *(to be implemented in April 2014)* | Absolute energy limits for space heating introduced in 2013 version (domestic only) | kWh/m²/yr |
| | Primary energy limits | kWh/m²/yr |
| ***Energy Performance of Buildings Directive (EPBD)*** | Energy Performance Certificates<br>Display Energy Certificates | A–G rating on a 0–100 scale based on $CO_2$ emissions<br>A–G rating on a 0–100 scale $CO_2$ emissions<br>kWh/m²/yr for space heating and electricity in a table at the bottom |

| Name of standard, voluntary or mandatory | Form of rating or compliance target | Measure |
|---|---|---|
| *BREEAM for non-domestic buildings (Code for Sustainable Homes for the domestic sector)* | Design and assessment methodology applied across a range of sustainability measures from energy to ecology incorporating Building Regulations $CO_2$ and energy use | Credits awarded against each factor and then multiplied using a weighting system to deliver a single score from Pass to Outstanding |

Standards expressed in carbon dioxide emissions can distract designers from delivering low energy outcomes, focusing their intention instead on usually costlier low or zero carbon technologies. Carbon emissions are intangible, invisible and not directly measurable; they need to be calculated using factors based on a range of assumptions that are also liable to change as insights develop.

More importantly, reducing carbon emissions by using low carbon technologies does not necessarily save energy (or energy costs); indeed, it may increase operating costs, whereas saving energy will generally reduce running costs. The wording of the Energy Performance Certificate also adds to the confusion where it states against the grade marker: "This is how energy efficient the building is". Well, only if the building performs as designed... More on that later.

The way for us to reduce carbon emissions most effectively is to use the lean, mean, green hierarchy.

**FIGURE 1: The lean mean green hierarchy**

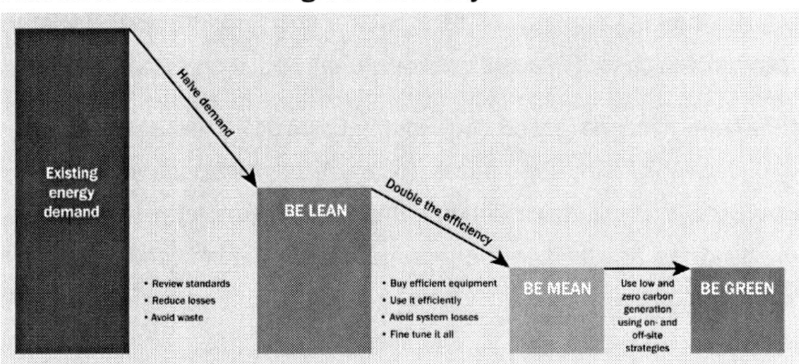


**SOURCE:** Courtesy Bill Bordass

## Relative versus absolute targets

The regulatory framework has generally used *relative* targets or percentage reductions. The starting point is the Target Emissions Rate or TER which is the $CO_2$ emissions of a notional building of the same dimensions designed to comply with 2010 building regulations. The Building Emissions rate or BER is the 'as designed' emissions of the real building, designed to comply with or exceed building regulations. Expressing BER as a percentage of the TER gives a figure for the relative 'goodness' of the design in terms of regulated energy use (those energy uses associated directly with the building rather than the discretionary activities of the occupants). So the first problem is that this percentage does not provide an instinctive sense of absolute energy performance and the second problem is that both the TER and BER are theoretical figures, often bearing little resemblance to the intrinsic energy performance of the physical building. So, in theory carbon emissions may be reduced by 44%. But this is no more than a theory.

*If you don't know where you want to go, all roads are equally good...*

It is more robust and much clearer to take energy values as the primary target, which can subsequently be measured, both overall and by end-use.

How many miles per gallon does your building do? Using a simple, easy-to-understand metric like mpg that focuses attention on the final outcome would make a major contribution to improving the energy performance of our buildings. An *absolute* energy target defines the intended outcome in terms that are directly measurable. It provides a clear, understandable goal for energy outcomes for the client, the building manager, and its occupants.

This is not to argue for dropping carbon emissions as an indicator. Clearly reducing emissions is important. But any single indicator will not be adequate as each can create unintended consequences of one kind or other.

The Display Energy Certificate has separate thermal and electrical indicators and a combined indicator, which with UK policy has tended to be $CO_2$, and in other countries often primary and/or delivered energy. Indeed, the Energy Performance of Buildings Directive EPBD requires the use of a primary energy indicator in national building regulations and this is now being implemented in the UK (although the European Commission has instigated formal proceedings against the UK regarding failure to put the directive fully into national law). A primary energy indicator has the effect of placing a cap on grid electricity and requires an overall focus on energy efficiency.

What this all illustrates is that there is no substitute for simple metrics based on measurable outcomes.

kWh/m$^2$/yr space heating

kWh/m$^2$/yr electricity

kg $CO_2$/m$^2$/yr

So how many kWh/m$^2$/yr does a low energy building use?

## Embodied energy and carbon

There is also often confusion between operational energy and carbon and embodied energy and carbon. Operational carbon arises from the energy used to operate a building. Embodied carbon is a measure of the energy used and carbon emitted to make the products that are used to construct a building.

Striking the right balance can be difficult. For example, for each tonne of cement that is manufactured, a tonne of $CO_2$ is produced. However, concrete buildings can deliver very low energy (in-use) performance because they provide thermal mass. When it comes to mechanical and electrical plant and equipment, the embodied carbon and energy can be even higher (compounded by the fact that manufactured plant has a relatively low materials conversion factor – only about 10% of the primary material required to manufacture a piece of equipment ends up in the installed device).

If the carbon emissions related to operational energy use are high, the proportion related to embodied carbon is likely to be small. As the operational carbon emissions reduce, so the proportion related to embodied carbon increases. A study by NHBC comparing the overall emissions from a timber-frame and concrete-block house illustrated that, ultimately, the operational emissions dominate and the key focus should be on building with the minimum use of resources, whatever the materials chosen may be.

Furthermore, those readers of this book concerned with existing buildings will have little opportunity to affect the embodied carbon other than when undertaking retrofit measures.

Broadly speaking there will be a trade-off between embodied and operational carbon. A well-designed building can aim to minimise both.

**Key Learning Points**

- It is vital to be clear about the distinction between energy and carbon to make sound investment decisions.
- Reducing energy use will generally be the cheapest way of reducing carbon emissions.
- The regulatory framework is largely based on percentage reductions in carbon emissions and building performance is based on theoretical rather than actual figures.
- The use of sustainability methodologies is no guarantee of obtaining a low energy building.
- Using absolute energy and carbon targets expressed in kWh/m$^2$/yr and kg $CO_2$ /m$^2$/yr defines intended outcomes in terms that are directly measurable.
- Operational carbon arises from the energy used to operate a building.

- Embodied carbon is a measure of the energy used and carbon emitted to make the products that are used to construct a building.

PART 2

# The Performance Gap

## Buildings often use much more energy than expected

*"We know there are known unknowns; that is to say we know there are some things we do not know. But there are also unknown unknowns – the ones we don't know we don't know."*
DONALD RUMSFELD, FORMER SECRETARY OF STATE FOR DEFENSE, UNITED STATES

**IT WOULD BE UNFAIR TO COMPARE** the UK buildings sector with the man who used this piece of obfuscation to justify going to war on a false premise. But there are parallels with the buildings sector, not only in the UK, but worldwide.

Awarding recognition for outstanding performance is a feature of most industries – usually an opportunity to hobnob with one's peers and eat expensive dinners. Architecture is no exception and some of the most public, high-profile awards are for buildings. Buildings are highly visible, provide great images for the press, and will be in existence for many, perhaps hundreds, of years.

The kinds of buildings that win prizes tend to be prestigious, iconic, and maybe award their designers for their chutzpah – being the tallest (a never-ending race apparently), changing the skyline of a major city, making the building an attraction in and for itself, never mind what it contains. Strange shapes seem to work a treat.

The fame awarded these kinds of buildings is rarely overshadowed by subsequent criticism from, for example, their tenants who discover that the usable space is heavily constrained because the shape does not lend itself to installing office furniture efficiently. Or complaints about frying in summer and freezing in winter.

> ***"On the day we moved in, every member of staff was given a new computer in a cardboard box. It was high summer. By the end of the week the building looked as if it were in a Third World country, with flattened cardboard boxes hanging at all of the windows on the south-facing side. In the winter, because the secondary heating system had been omitted to save costs, we were all freezing and people smuggled in their own fan heaters."***
> **LONG-SUFFERING OCCUPANT OF A NEW UNIVERSITY BUILDING**

In effect, buildings have been celebrated for their aesthetics and engineering ingenuity, but rarely for how they perform in use. Over the years, some design teams have made proposals to look at in-use performance as part of their service to clients, but clients have been reluctant to pay for it. Failure to evaluate performance and provide feedback means that the same problems have been able to persist for decades.

During the 1980s, sustainability rose up the agenda, and the buildings industry was suddenly being asked to respond to pressure for radically improved product quality, business performance *and* sustainability: the triple-bottom line required buildings and the management of them to deliver simultaneously social, economic, and environmental benefits.

This is when we entered the new phase of giving awards to 'green' buildings. The sustainability was usually self-evident, based mainly on visible features like solar thermal panels, wind turbines, solar photovoltaic panels, fuel cells, green roofs – a lot of expensive kit which indicated 'green-ness' that sadly didn't necessarily convert into improved energy performance, or even improved carbon performance. But as we will learn later, the best low energy buildings have characteristics that are largely invisible.....and have sadly therefore not won 'green' building awards.

## Establishing real energy outcomes

In the mid- to late 1990s the PROBE studies[6], were undertaken to study in depth the performance in-use of a range of building types. Sixteen buildings were evaluated for technical performance, energy performance, and occupant and management satisfaction. The confidential technical report on each building was then converted by *Building Services Journal* into an article of typically five or six pages. (Several other similar studies have since been added to the collection).

*"Most of the buildings claimed to be energy efficient. The energy performance was sometimes very good but more often disappointing, and usability and manageability often left much to be desired. But the range of annual consumption and emissions was massive, with a factor of six between the highest and the lowest, and even greater variations in some individual end uses."*[7] There were a few very good buildings so that lessons were there to be learned about what does work as well as what doesn't.

Numerous reasons were identified as leading to the problems that arose. In essence, it became clear that design-stage energy estimates could not be taken at face value, and in-use performance would have to be measured to determine actual energy use and understand what led to it. Then the multiple issues which gave rise to the 'performance gap' would need to be analysed one-by-one – and resolved (if possible).

Over a decade later, in 2010, with the PROBE studies remaining almost unique, the Technology Strategy Board (TSB) launched its Building Performance Evaluation (BPE) programme, providing funds to organisations to undertake post-construction and in-use performance evaluation either of buildings in the completion and hand-over phase or those in the first three years of their life. Since then, over one hundred studies have been put in train.

To date, few results from the studies have been made available, though a series of reports are complete on many of the buildings. So what are we waiting for?

The fact is, many of the results are proving sensitive (if potentially extremely valuable to the industry) and in some cases there is reluctance to go public if the studies show that the buildings are not performing as expected. More may prove willing when there are enough similar examples to confirm that they are not alone. Organisations such as the National Energy Foundation are hopeful that further examination of findings in confidence is likely to enable lessons to be extracted and shared, if not brought to life through being illustrated by practical examples. Some (partial) results have been published on CarbonBuzz[8], an on-line platform for benchmarking and tracking energy use and carbon emissions in projects from design to operation. CarbonBuzz is intended to encourage users to go beyond compliance with mandatory Building Regulations calculations and to refine estimates to account for additional energy loads in-use. The platform allows users to compare design energy use with actual energy use so as to help users close the performance gap.

CarbonBuzz allows data to be uploaded on an anonymous basis, but those who are willing to reveal the results of their monitoring and measurement can also publish them by name. Further results for the non-domestic buildings in the BPE competition will be uploaded onto CarbonBuzz and those for the domestic sector will be made available on EMBED, the Energy Monitoring and Building Evaluation Database web application funded by Technology Strategy Board[9].

Herein lies a real dilemma. The culture of the buildings industry has been, in the first instance, not to measure at all and contractual arrangements have avoided attributing the achievement of in-use energy performance to any part of the supply chain. Now that measurement is revealing serious issues with building performance, there is understandable reluctance to risk ramifications and hence we risk losing the potential to learn the lessons and improve outcomes.

Tools like CarbonBuzz contain the benchmarks published by CIBSE in TM46[10] which allow you to compare your own buildings one with another, and to compare your building(s) with those of others of a similar type. However, our failure to invest in monitoring and measurement means that these benchmarks are not up-to-date or necessarily accurate.

The few companies that are making their data public are to be applauded – and thanked. It is clear that the intentions of design and construction teams are good – no-one deliberately delivers a bad building. But good intentions based on ill-defined objectives, misconceptions, and unsound evidence create major obstacles to success.

Until the industry as a whole learns transparency, we will be unable to resolve the issue of the performance gap and will be trapped in that murky world of Donald Rumsfeld and the known and unknown unknowns. The power to make this change is primarily in the hands of the construction client – the more we ask the question and the more knowledgeable we are in interpreting the answer, the better our buildings will become.

**Key Learning Points**

- Buildings which win awards for their 'sustainability' are no more likely to be low energy than any other building.
- Very few buildings in the UK perform in line with their design intent.
- Making building energy performance visible is moving onto the agenda although there is still a long way to go.
- There is a need for significant investment in building benchmarks.

PART 3

# Why Do We Have the Performance Gap?

## Everyone's playing the game – but will it lead to a good building?

**DO YOU REMEMBER PLAYING PICTURE CONSEQUENCES?** Every player first draws the feet of an animal or person, and then folds the paper to hide the drawing and passes it to the next person; then everyone draws a set of legs, and then a body, and finally a head. Usually the result was very amusing.

The building process is not so different from drawing an animal using picture consequences. Everyone knows that they are aiming to create a building; but with little idea of the exact outcome they are aiming for, each makes a contribution that may or may not correspond with what others involved are doing. We often end up with a very strange beast...

Portcullis House, an office building in Westminster, London, was commissioned in 1992 and opened in 2001 to provide offices for 213 Members of Parliament and their staff. It was hailed as a flagship BREEAM Excellent building, intended to be "ultra low energy ". Its Display Energy Certificate rating of G demonstrated the reality. (Subsequent energy management measures have made it an E.)

**FIGURE 2: Portcullis House**

## There are a great many participants in the buildings cycle

One of the reasons for this is that delivering buildings is complex and the buildings cycle involves a great many players, each of whom can inadvertently confound the intent of delivering a building with excellent energy and carbon performance.

Figure 3 lists just some of them in five principal stages of the buildings cycle – acquisition, specification and preliminary design; detailed design and construction; commissioning, verification and handover; marketing and sales; occupation, management and maintenance. Obviously, many of these stages run concurrently but to analyse them we need to separate them. And all of them operate within and interact with the policy and regulatory framework and the property market.

**FIGURE 3: The Buildings Cycle – phases and roles**

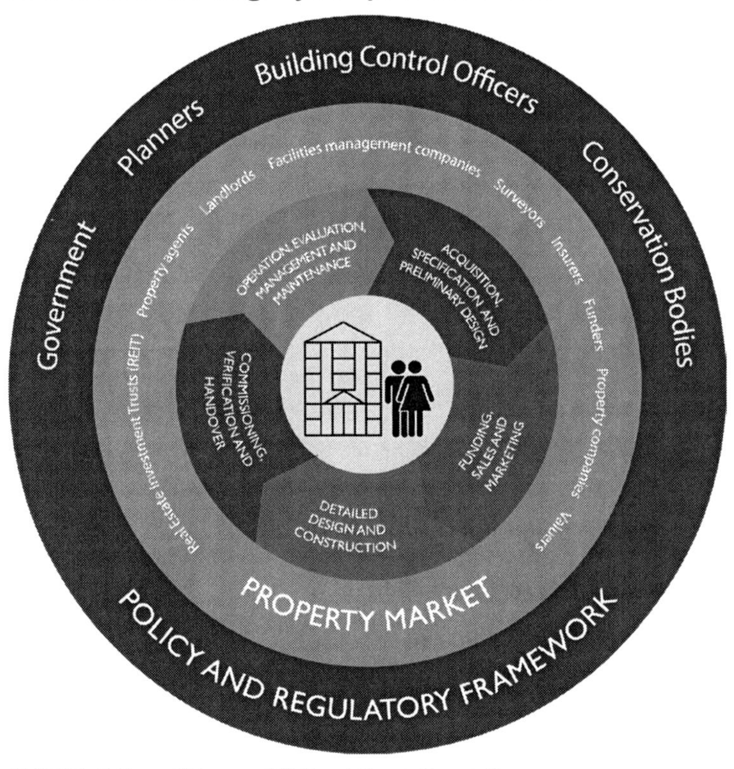


**SOURCE:** © Green Stripes and National Energy Foundation

The typical phases of a building, with the main participants and processes are listed below. Some of the sub-phases – those that are italicised – reflect processes that should be followed, although as demonstrated later, are often overlooked.

1. **Acquisition, specification and preliminary design** *(as per the RIBA Plan of Work Stage A)*

- government sets the regulatory framework
- planning, conservation and building control departments *check compliance*
- the client (of which there are many different types) decides to acquire a new building or to refurbish an existing one, specifying what they are after, either on a speculative basis or for their own future occupation
- the speculative developer takes a view on the market and the possible rate of return – one that must satisfy their shareholders
- a public or private sector organisation decides to invest in a new or existing building or buildings for its own use or for that of its stakeholders (offices, schools, hospitals, residential social landlords)

2. **Funding, sales and marketing**
    - the property is financed
    - the property is marketed
    - a lease is signed
3. **Detailed design and construction**
    - an architect (and architectural technologist) is appointed alongside a building services engineer, a structural engineer, and a quantity surveyor and a sustainability consultant;
    - the architect has to consult the planners and come up with a design that obtains planning approval, including, for example, satisfying transport and conservation concerns;

- the project may then be handed over to another architect and specialist design team for detailed design, potentially as part of a design and build contract
- the full specification is drawn up, costed and priced;
- the contractor is appointed;
- suppliers are asked for the specification of their material, components or equipment;
- the property is constructed;
- material, components and equipment are installed.

4. **Commissioning, verification, handover**
   - *components and equipment are commissioned;*
   - *verification is undertaken – have you got what you asked for?*
   - the property is handed over.
5. **Occupation, *evaluation*, management and maintenance**
   - the property is occupied
   - the facilities are managed
   - *the building is evaluated.*

## The different players have different priorities

Figure 4 shows the results of a Building User Survey (BUS) of a non-domestic building. (Over 800 of these have been completed in sixteen countries over the last fifteen years)[11]. Using a BUS, the occupants rate the building from the key perspectives that make it comfortable and usable.

**FIGURE 4: Results of a building use survey**

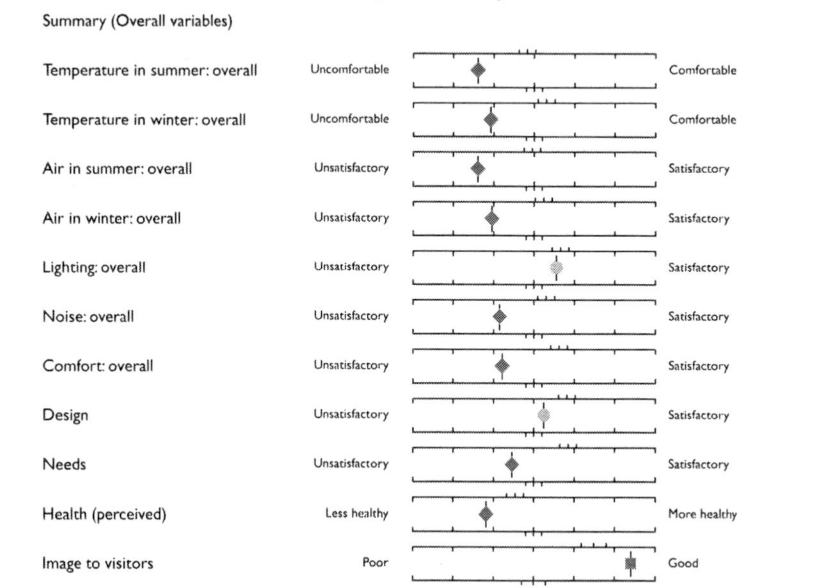


This particular building – which must remain anonymous – has what has proved often to be a characteristic signature of those buildings studied – a high rating for 'Image to visitors' but poor or very poor ratings for comfort and satisfaction variables.

Equally, Figure 5 shows the size of the energy performance gap – the difference between the asset rating based on the modelled design and the actual operational energy based on measured use.

**FIGURE 5: The energy performance gap – the main contributing factors**

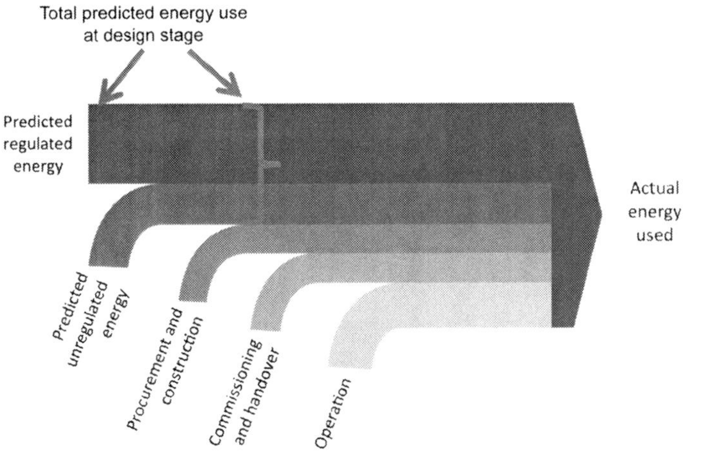

Total predicted energy use at design stage

Predicted regulated energy

Actual energy used

Predicted unregulated energy

Procurement and construction

Commissioning and handover

Operation

**SOURCE:** © National Energy Foundation

How does this happen? The table below summarises common problems that arise with key players in the buildings cycle with respect to a building's energy and carbon performance.

**TABLE 2: Roles in the buildings cycle and common problems in relation to building energy performance**

| ROLE | COMMON PROBLEMS |
|---|---|
| Planner | • Is unlikely to have an understanding of energy and buildings.<br>• Is likely to use inappropriate mechanisms for delivering greener buildings such as demanding the installation of large amounts of renewable kit, or relies on published codes and benchmarking tools without understanding the performance gap.<br>• Is driven by planning guidance with no allowance for the impact on building performance. |

| ROLE | COMMON PROBLEMS |
| --- | --- |
| Building control | • Primary focus is health and safety and examining plans, calculations and building works for compliance with the Building Regulations.<br>• Heavily reliant on visual inspection of plans and site works.<br>• Unlikely to have expertise in assessing the energy impacts of design and construction. |
| Property companies | • Those responsible for marketing and renting properties rarely make a nod to energy performance, even when there is a good story to tell.<br>• 'Grade A' invariably means atrium, air-conditioning, lifts, focusing on attributes that customers are in the habit of looking for – prestige, useable space, location, maybe insurance consequences (risk) but not usually energy, comfort, or occupant satisfaction and productivity.<br>• Not a priority nor high profile though EPBD is taking us in that direction as will straightforward business concerns. |
| Client/ funder | • Allows capital budget to act as the overall arbiter of design decisions without exploring how different performance options could fit into the same financial envelope.<br>• Applies upfront capital cost rather than whole-life costing.<br>• Thinks inputs rather than outcomes.<br>• Doesn't have standard metrics to apply when specifying a building in terms of its performance.<br>• Relies on the design team for advice (see below).<br>• Believes that using sustainability 'benchmarking tools' is a guarantee of delivering buildings with sustainable performance in use. |

| ROLE | COMMON PROBLEMS |
|---|---|
| | • Does not include occupants at the specification stage and is not aware of the significant impact on energy performance that will arise from changing user specifications, such as hours of operation. Does not invest in building/facilities managers at a level appropriate to the complexity of the building and its systems.<br>• Does not expect to have to evaluate whether its money has been well spent.<br>• Is unwilling to pay for building performance evaluation or to establish whether its occupants like the building they are in. |
| Architect | • Is unlikely to have had significant training in energy and buildings.<br>• May be attracted by green style over substance.<br>• May be taken off the job once the aesthetics have got the building through planning.<br><br>The walkie-talkie building at 20 Fenchurch Street (London) whose concave profile was channelling the sun's rays into a concentrated beam onto Eastcheap, which was apparently able to blister paint and fry an egg.<br>'Death-ray' designer Rafael Viñoly, visiting London this week, said the original design of the buildings had featured horizontal sun louvres on the south-facing façade, but these are believed to have been removed during cost-cutting as the project was delivered. ***"One problem that happens in this town, is the super-abundance of consultancies and sub-consultancies that dilute the responsibility of the designer to the point that you just don't know where you are anymore."***[12] |

| ROLE | COMMON PROBLEMS |
| --- | --- |
| | • Relies on the sustainability consultant to do the 'energy bit'.<br>• Is likely to leave the detailed design to architectural technologists who may have to draw architectural details to deliver features which are fundamentally unbuildable.<br><br>***"If we're supposed to be delivering a low energy building, one of the first things we have to do is go through the architect's detailed plans and ensure that they have actually drawn what's needed correctly. More often than not, they haven't. But getting those details right from the outset is essential."***<br>**SITE MANAGER, SPECIALIST REFURBISHMENT CONTRACTOR** |
| Sustainability consultant | • May have knowledge of a range of sustainability issues but often has no special expertise in energy and buildings.<br>• Uses the National Calculation Methodology (SAP for domestic and SBEM for non-domestic buildings) with all their simplifying assumptions and omissions (only certain energy uses are 'counted' for regulatory purposes and unregulated energy is ignored).<br>• Is not involved in the process of tendering and then delivery.<br>• Can be 'cut out' of the process as a way of reducing costs. |

| ROLE | COMMON PROBLEMS |
| --- | --- |
| Building services engineer | • Building services engineers are responsible for the design, installation, operation and monitoring of the mechanical, electrical and other systems required for the operation of modern buildings. This does not mean that they have any control over the design of the building fabric or its integration with services. |

**FIGURE 6: Palestra House, London**

The owners of this building in Southwark were so proud of its fuel cell Combined Cooling and Heating Power unit, that they created a special ground floor glazed box to make it visible to the public.

Unfortunately, its Display Energy Certificate rates the building as a G, with space heating energy use nearly three times that for a typical building of its type.

It so happens that the building is directly opposite a tube station where, helpfully, the lights illuminating the concrete surface below them shine day and night.

**FIGURE 7: Southwark tube station**

• Is likely to oversize plant to try and minimise future complaints about lack of warmth or coolth.

| ROLE | COMMON PROBLEMS |
|---|---|
| | • Likely to be contracted with fees linked to the overall value of the construction cost installed with no incentive to minimise active plant and equipment.<br>• Not included or equipped to participate in early stage development of building energy strategy. |
| **Structural engineer** | • The main objective of the structural engineer is to ensure that the building doesn't fall down. This means that they can decide to use and/or add extra (and quite possibly unnecessary) structural elements for strength that undermine the integrity of the thermal envelope. |
| **Contractor project manager** | • The role of construction contractors is to oversee and manage projects, ensure the work meets client specifications and building codes and, most importantly, comes in within the contract cost.<br>• If the contractor is not committed to delivering a building to a given in-use energy target, the contractor is likely to be responsible in many ways for compromising performance.<br>• Specification may drift where the words 'or equivalent' are not understood or checked so that materials or equipment can be substituted with less suitable or effective alternatives.<br>• Major contractors employ professional teams, but rarely employ trades staff, taking on a wide variety of sub-contractors, frequently of a number of different nationalities with varying English-language skills. These sub-contractors may change from week to week.<br>• Those on site at any time are unlikely to have a good understanding of the significance of their work to future building energy use or to have any mechanism to work together to achieve good performance. Expecting them to deliver, for example, an airtight building, would require frequent training of new staff and constant monitoring of their work. |

| ROLE | COMMON PROBLEMS |
|---|---|
| Quantity surveyor | • Is responsible for cost planning and commercial management during the entire life cycle of the project from inception to completion. They are colloquially known as 'cost consultants'.<br>• They are also involved in 'value engineering', an approach which aims to provide the necessary functions in a project at the lowest cost. Where the energy performance outcome of a building is not well understood, value engineering is frequently the cause of a significant loss of energy performance in terms of the fabric, the services, and the ability to monitor actual energy use.<br><br>***"When I went round the building with the architect, so much was different about it, I had to ask her whether this was really the building for which they had submitted their application. All the sub-metering control systems were gone for a start – everything that we would need to establish the actual performance of the building in operation!"***<br>**BUILDING PERFORMANCE EVALUATOR** |

| ROLE | COMMON PROBLEMS |
|---|---|
| On-site workers | • There is a multiplicity of trades, most of whom are skilled in only one function.<br>• High turnover on site with training having to be repeated on a weekly basis.<br>• Multi-lingual taskforce, with common language often not English. Often paid on piece-rates.<br>• Trades tend to be single task-focused, and work in silos. 11% of manual workers have no qualifications. Can be asked to work out of sequence to meet tight deadlines. Very few understand energy and buildings. Probably have an unwarranted confidence in the long-term efficacy of silicon sealants.<br><br>***"One lad was looking hard at how to fit the window into the section of wall. "Get a move on", said the lecturer. "I'm thinking about the next bloke and what he'll have to do." Lecturer: "It's not your f***ing job to think about the next bloke!""***<br>EXTERNAL TRAINER ON HOW TO DELIVER AIRTIGHT BUILDINGS ON A CONSTRUCTION COURSE IN A FURTHER EDUCATION COLLEGE, REPORTING ON THE TRAINING ACTIVITY THAT INVOLVED A PRACTICAL EXERCISE, DELIVERED WITH A COLLEGE LECTURER IN ATTENDANCE. |

| ROLE | COMMON PROBLEMS |
|---|---|
| Suppliers of material, compo-nents or equipment | • Design teams and contractors will rely on the specifications provided by their suppliers to deliver the functionality to which their design and contract commit them. |

**THE SHARD**

**Property Type:** Office

**Location:** The South Bank of the Thames has always played a significant role in the rise of London as a world city – an economic, social and cultural magnet that draws people to live and work in the area.

London Bridge Quarter stands at the heart of this vital community. The architectural masterworks of The Shard and The Place, the transformed transport hub, the new retail space and the landscaped public realm reveal the city's confidence and capacity for reinvention.

**Specification:**

- 2.7m floor to ceiling height, 6.25m in Entrance Hall
- 4 pipe fan coil air conditioning
- Triple skin facade with automatic blinds
- 1.5m planning module
- 150mm raised floors (overall)
- Designed to achieve «Excellent» Breeam rating
- Floor loading of 3.5kN/m² + 1kN/m² (+7.5kN2 in part)
- Nine 21 person double destination hall-call passenger lifts serve the office floors
- All office floors serviced by two goods lifts (3,500Kg each)
- 155 bicycle spaces with showers & lockers on Level 0
- 48 car spaces

- Manufacturer performance declarations, which are legally required to be in conformity with harmonised European Standard, do not provide adequate 'in-situ' performance values for input into energy modelling software.

| ROLE | COMMON PROBLEMS |
|---|---|
| | • Products and materials are generally tested in isolation, as individual components, not as the systems or fabric assemblies constructed on site.<br>• Manufacturers often provide insufficient guidance on product installation and commissioning for site teams.<br>• Many are normally working at the end of the contract period and are rushed to complete their work for building handover meaning that plant and equipment commissioning is omitted. |
| Building and facilities managers | • These are the people with the hardest job – given a building to manage, quite often with limited resources and an introduction to the building fabric, services and control systems that lasts an hour if they're lucky. Likely to have many other responsibilities too. It is common for air-conditioning and heating systems to be running in competition with one another, with no-one noticing.<br>***"The air-conditioning system in the pictures room was designed to maintain them at a constant temperature. When it broke down, the temperature rose by a couple of degrees. It didn't matter though – the temperature remained constant – which is all that matters. The University decided to leave well alone – and the museum's energy bills went down significantly."***<br>**CURATOR, UNIVERSITY MUSEUM** |
| Occupiers | • These are the people who feel hot when others feel cold, need the windows open when others need them shut, and leave their computers and lights on all night.<br>• They often have to wear jackets in the summer because the air-conditioning system maintains the temperature too low. |

| ROLE | COMMON PROBLEMS |
|---|---|
| All players | • Generally find themselves working independently of other parts of the team.<br>• Lack of integrated design and construction processes make the delivery of performance outcomes difficult to achieve.<br>• There are few formalised feedback mechanisms between design, procurement, and construction teams and final building managers and occupiers.<br>• There is limited information about in-use building performance with a consequent lack of evidence-based benchmarks to inform the design and construction of the next property.<br>*"I went to Germany to watch a Passivhaus home being built. Six guys, multi-tasking for three weeks, and with the sole responsibility for constructing an air-tight shell, built the shell from excavations to window installation. Then I came home and looked at my 32 trades on site, each of them eating their lunch in their own white vans..."*<br>DIRECTOR, MEDIUM-SIZED BUILDING COMPANY |

## The regulatory modelling tools may hinder rather than help

Even those that have been trying to deliver genuinely low energy buildings have been working with one hand tied behind their back.

Part L of the Building Regulations sets standards for space and water heating and fixed lighting, but excludes equipment and appliances and plugged-in lighting – on the basis that the amount of energy they consume is influenced by the building's occupants. This means that buildings contain 'regulated' and 'unregulated' loads. So when designers think that they have calculated the predicted energy use accurately, they have actually omitted significant energy uses and impacts on energy uses.

The National Calculation Methodology tools of SAP and SBEM, have been designed to demonstrate compliance with Building Regulations. Although the basic model on which the tools are based is grounded in building physics, they have not been validated by real data from buildings in use since their development in the early 1980s. They also, necessarily, contain a series of simplifying assumptions, which diminish their accuracy for the particular building being designed.

As these calculations are usually undertaken by a 'sustainability consultant', one of whose tasks is to help the designer reach compliance, the tool then starts to be used to identify design solutions – a purpose for which it is neither designed nor intended.

Figure 8 below illustrates the estimated size of the performance gap in a non-domestic building. It compares the calculated Building Emissions Rate at the design stage pre-handover, with that at post-handover, and then again with what actually happens to energy use once the building is operational. Even before the building has been occupied, its predicted energy use has doubled. And it more than doubles again once the building is in use.

**FIGURE 8: Analysis of the performance gap**

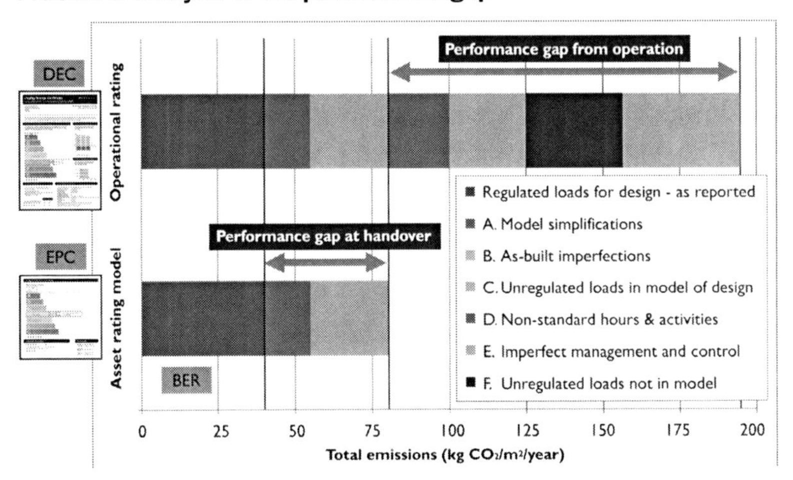


**SOURCE:** Courtesy Robert Cohen, Verco: Closing the Performance Gap between the Predicted and Actual Energy Performance of Non-Domestic Buildings, paper for the Green Construction Board Buildings Working Group

## Buildings are an energy system where energy is integral, not a separate 'sustainability' issue

Two further issues arise from this set of approaches.

First, divorcing the fabric from the other energy uses in a building is artificial in that a building is a system. For example, the heat required from a building's space heating system will differ according to how much solar warmth it gains, or how many people are occupying the building and the activity they are involved in; every person sitting in a building emits around 100W. Electrical equipment, lights and appliances all give off heat in use.

Server rooms often generate large amounts of heat which then require cooling equipment – while the space heating systems are delivering heat to the remainder of the building. An energy-efficient building is one that delivers the balance of energy in and energy out at the lowest input energy demand for a building with comfortable conditions.

Second, the very existence of 'the sustainability consultant' and the notion of adding 'sustainability features' to a building makes it all too easy to cut them out when looking to save costs. This is why sustainability needs to be fully integrated into the building design, rather than treated as an add-on, always vulnerable to being cut off.

**Key Learning Points**

- The state of knowledge and understanding about building energy use and carbon emissions of the buildings sector in general is parlous.
- Many of the tools that should improve building energy knowledge mislead rather than inform.
- In general, the construction industry is seriously dysfunctional.
- The structure of building procurement, the adversarial and blaming industry culture and systems all conspire against the delivery of better buildings and are in need of root-and-branch reform.
- The buildings sector – occupants included – will have to overcome a series of tough challenges if we are to start delivering and improving buildings to achieve excellent energy and carbon performance.

PART 4

# Where Do We Go From Here?

## Effective solutions are available

**WITH SUCH A STRING OF ISSUES TO ADDRESS**, the task of significantly improving outcomes can seem insurmountable.

But never fear, help is at hand! You can get improved energy performance from your buildings by applying the following seven simple steps.

**Ten steps to radical change**

1. Talk a common language with everyone on the team.
2. Focus on **outcomes** rather than inputs.
3. Learn some building physics – the basics are set out below.
4. Take the 'fabric first' approach: at the heart of a low energy building is nothing more complicated than a well-designed and executed building envelope.
5. Apply the same approach to building services.
6. Understand the relationship between fabric, building services and their occupants.
7. Learn to count and add up: it is possible to establish realistic figures for actual energy use simply by learning to count and add up.
8. Use an integrated process.

9. Verify you've got what you asked for – throughout the construction process for new or refurbished or using DECs and EPCs for existing buildings.
10. Evaluate building performance: measuring and monitoring outcomes is the best way to diagnose problems, and celebrate successes.

**FIGURE 9: Closing the feedback loop**

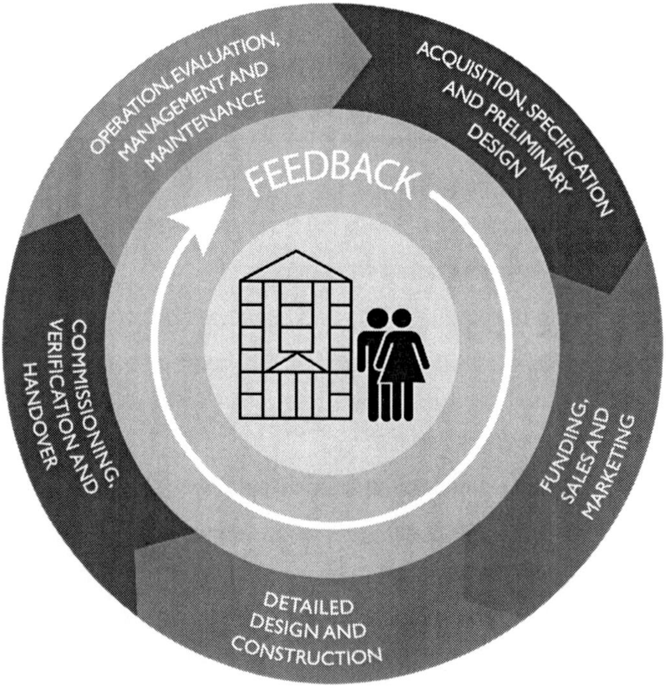

**SOURCE:** © Green Stripes and National Energy Foundation

The following section illustrates each of these points.

1. **Talk a common language**

Where you are about to build a new building, or to lease, rent or refurbish an existing building, establish your absolute energy and $CO_2$ emissions targets at the outset. Ensure that the targets, and any contributory targets, are the common currency for funders, all members of any design team, sales and marketing teams, existing or proposed occupants, and building managers.

**Get everyone to read this book so that they understand the challenges**

2. **Focus on outcomes**

Ensure that your procurement processes focus on the target outcomes, rather than inputs. Make sure that every proposed change to the design, to construction materials, components and detailing, and to equipment and appliances is analysed for its impact on the intended outcome. Contract for your target outcome and incentivise everyone in the supply chain to deliver their contribution to it, with penalties for failure if need be.

Your procurement contract should include the continuing involvement of design and construction teams for two years beyond practical completion in order to track and rectify any problems with initial systems. This has budgetary implications, and requires the appointment of teams willing to be involved in finding out what works and what does not in their building. But any upfront costs should be well re-paid in terms of getting a better building at the end of the process.

Use of Building Information Management (BIM), integrated with a robust energy modelling system, should help with the delivery of the target outcomes.

Setting aspirational and measurable targets need not apply to energy outcomes alone. The following report illustrates the impact of setting – and not setting – an airtightness specification.

*The Elizabeth Fry Building (EFry) was initially pressure tested by the BSRIA Fan Rover in December 1994 to check that the airtightness met the performance criterion for the building. The building was required not to exceed 1 ac/h (air changes per hour) at a test pressure of 50 Pa. The test result then was 0.97 ac/h @ 50 Pa (equivalent to 4·2 m³/h/m² of envelope area) which met the criterion.*

*The BSRIA has [in the four years since] undertaken measurements on 14 office buildings which have been constructed with an airtightness specification of less than 5 m³/h/m² @ 50 Pa. The average value for all was 5.7 m³/h/m², while the average value for the top ten of these buildings was 4.1 m³/h/m², the same as the airtightness value achieved by the Elizabeth Fry Building back in 1994.*

*The average for office buildings tested by the BSRIA* ***without an airtightness specification*** *(my emphasis) was 21.8 m³/h/m², with the worst being an air-conditioned office building at 40.1 m³/h/m². The best office building tested was 2.78 m³/h/m², and the best superstore was 1.65 m³/h/m².*[13]

### 3. **Learn some basic building physics**

Understanding the principles of low energy buildings is the same as knowing how to keep warm – take shelter, wear plenty of warm layers, held close to stop draughts, and drink hot soup! To keep cool in summer, keep out of the sun and wear loose clothing that creates draughts as you move.

Key is to appreciate that buildings are energy systems composed of interactions between the building's external environment, the building itself (orientation, form and fabric), its building services, and its occupants.

#### ***Environment***

It stands to reason that climate and location will impact on energy use – the greater the difference between external and required internal temperatures, the more energy will be required to maintain the required comfort conditions. Similarly, a building exposed to wind and rain will have to work harder.

#### ***Orientation***

Ideally, in temperate climates a building – or at least its glazing – is oriented to make the most of solar gain to obtain free heat and for daylight, to reduce the need for electric lighting. At the same time, the landscaping and/or building design needs to incorporate features – strategic planting or solar shading – to protect its occupants from overheating. The leaves of deciduous trees can provide shade in the summer, while bare branches in the winter allow a bit more light in. The main priority is to stop the heat from entering the building so external shading or thermal shutters are more effective than internal blinds.

***Form***

Simple forms are more buildable and require the design and construction of fewer difficult junctions where heat loss is harder to prevent. Compact forms – squares and spheres – can be kept comfortable with less energy than buildings with lots of external surface area relative to internal volume.

***Lightweight versus heavyweight***

Lightweight buildings, such as timber-frame, heat up and cool down quickly. Heavyweight buildings with thermal mass, like *concrete have a slower thermal response.*

### 4. **Take the fabric first approach**

Heat loss is impacted by three main factors, usually in this order of significance – insulation thickness, air leakage and ventilation, and thermal bridging.

A low energy building fabric requires three things:

- a continuous layer of high quality insulating materials;
- a continuous airtightness barrier, that prevents air leakage and eradicates all thermal bypasses; and
- thermal bridge-free design (no construction element that compromises the insulating layer).

The insulation effectiveness of the different building fabric elements which make up the thermal envelope in terms of reducing heat loss is expressed in U-values. A U-value describes heat transfer through a building element. The lower the U-value, the lower the heat loss.

Air leakage is the uncontrolled flow of air through gaps and cracks in the fabric of a building (sometimes referred to as infiltration or draughts). This is not to be confused with ventilation, which is the controlled flow of air into and out of the building through purpose-built ventilators and ventilation systems that is required for the comfort and safety of the occupants.

Too much air leakage leads to unnecessary heat loss and discomfort of the occupants from cold draughts. The increasing need for higher energy efficiency in buildings and the need in future to demonstrate compliance with more stringent Building Regulations targets mean that airtightness has become a major performance indicator. The aim should be to 'build tight – ventilate right'. Taking this approach means that buildings cannot be too airtight; however, it is essential to ensure appropriate ventilation rates are delivered through intentional openings and mechanical ventilation.

Air leakage paths through the building fabric can be tortuous; gaps are often obscured by internal building finishes or external cladding. The only satisfactory way to show that the building fabric is reasonably airtight is to measure the leakiness of the building fabric as a whole. Air leakage is quantified as air permeability. This is the leakage of air ($m^3$/hr) in or out of the building, per square metre of building envelope at a reference pressure difference of 50 Pa ($m^2/m^3$/hr @50Pa). The alternative measure is the number of air changes per hour.

Typical UK buildings leak from most of their junctions (wall to window, wall to floor, etc.) and from service penetrations – where gas, water, electricity and telecommunications services pass through the fabric. Airtightness tests on new and refurbished buildings often reveal poor quality products (such as doors and windows), inadequate design and construction at junctions, and problems with installation.

Thermal bypass is heat loss created by convection within and between different thermal elements. An increase in heat loss of about 160% is not uncommon when air gaps exist behind the insulation from, for example, plasterboard and dab, carelessly filled cavities, or in party walls. Since 2010, party walls between two heated spaces have to be fully filled with edge sealing, but may work less well when one of two attached buildings is heated well and one is vacant.

Good airtightness is delivered with a range of building elements such as OSB board, plaster parging and finishing coats, glazing, concrete with special membrane tapes, mastics and grommets for sealing junctions and service penetrations.

If possible, specify air tightness testing to be carried out as soon as the building envelope is complete – i.e. before internal finishes are added. This makes it much easier to track down and solve leaks at source, rather than resort to the more superficial and often temporary sealing that is a typical response to air tightness deficiencies discovered when air tightness is tested for at completion.

Thermal bridges can be removed with careful design at minimal cost by, for example, continuing the thermal insulation below the floor level, or by ensuring that windows are installed in line with wall insulation, rather than sat on an external sill.

Gaps in an insulating layer are similar to a hole in a bucket – they breach the thermal integrity of the fabric and the building will 'leak heat'.

5. **Apply the same approach to building services**

Ensure that building services are also well-designed and executed. That means applying care to the specification, design, installation and commissioning of plant. Also ensure that occupants understand and are able to operate building services kit effectively.

6. **Understand the relationship between the building fabric, building services and the building's occupants**

***Ventilation***

An airtight building needs ventilation to ensure good indoor air quality. To comply with current Building Regulations, air permeability cannot exceed $10m^3/h/m^2$ @50Pa, although meeting the carbon target will generally require a maximum of $5m^3/h/m^2$ @50Pa. At this level, or not far below it, it is usually necessary to provide mechanical ventilation at least in the winter months - a mixed mode building. A building can be made more energy-efficient by installing mechanical ventilation with heat recovery (MVHR). MVHR recovers the heat in exhaust air drawn from kitchens and bathrooms to warm the incoming fresh air delivered throughout the building.

***Building services***

Building services are the 'active' aspects of a building. They include space and water heating; refrigeration and heat rejection; fans, pumps and controls; ventilation and humidification systems;lighting; office equipment; catering equipment; computer and communications rooms; telecoms; external lighting; security and lifts.

The length of the list alone is sufficient to indicate their complexity - to work well all of these services must relate to the overall operating strategy for the building and then be correctly designed, specified, procured, installed and commissioned in a manner that is consistent one with the other. Once a building is in use, services must be operated correctly, both to deliver comfort, usability and functionality to the occupants and to work in conjunction with each other in an efficient manner. The outcome of the use of services – comfort conditions in both summer and winter, good indoor air quality, and so on – are perceptible. But often those outcomes are delivered by the invisible elements of systems such as pumps and fans which use far more energy than expected.

In a building with a high quality fabric with a low heating demand, the energy usage of the building will be dominated by electricity and, in turn, be heavily dependent on occupancy and hours of operation, and all too frequently, systems that default to ON.

### *Operation, evaluation, management and maintenance*

> ***"The real performance of completed buildings depends on an intimate relationship between the properties of the fabric and services, the building's occupants and long-term management and maintenance activities. After all, it is people who create energy demand, not buildings."***
>
> **MALCOLM BELL, EMERITUS PROFESSOR OF SURVEYING AND SUSTAINABLE HOUSING AT LEEDS METROPOLITAN UNIVERSITY**[14]

There should be an appropriate match between a building's complexity – particularly its services – and the resource made available to manage it. Simpler buildings will intrinsically be easier to manage. Occupants should be an integral part of the energy management team, including those responsible for specifying new equipment, such as IT managers, who all too often increase loads by failing to take account of the energy demand of new kit.

Central, computer-controlled building management systems (BMS) have four basic functions: controlling, monitoring, optimizing and reporting on the building's facilities, mechanical, and electrical equipment to deliver comfort, safety, and efficiency. However, a BMS is not a panacea and can be subject to the same problems of incorrect commissioning and operation as the different plant and equipment it is there to manage. Good practice is now to re-commission the BMS on a six-monthly basis.

PROBE characterised four building types, recognising that all too often more technically complex buildings were not matched by the same level of building management input.

**FIGURE 10: Probe classification of building types and management input**

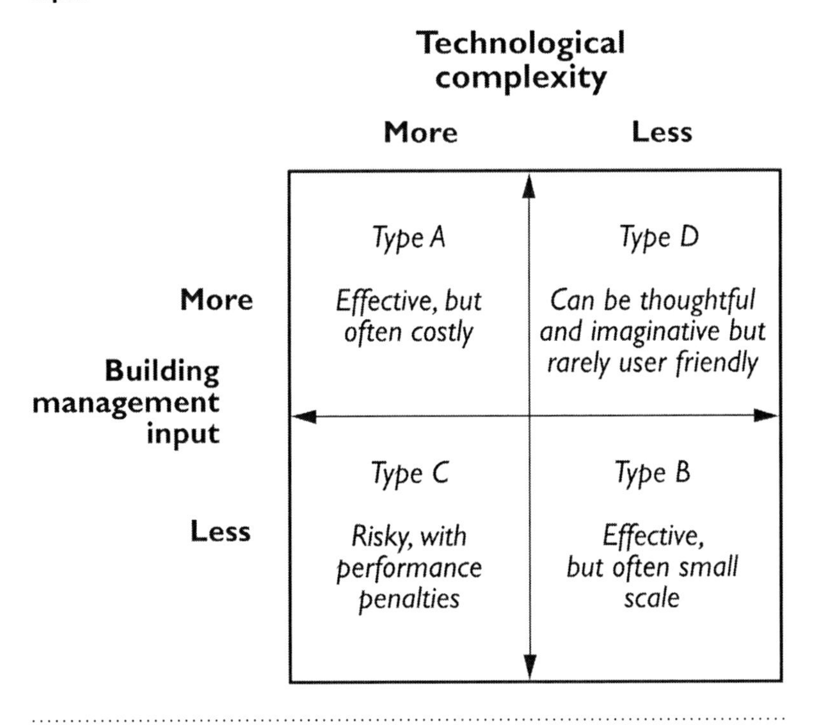

7. **Learn to count and add up – calculate how much energy you are using from basic data**

The PROBE Strategic Review quoted energy expert Amory Lovins who said *"much energy consumption comes from compounding of unnecessary loads."* It went on: *"Relatively inefficient services can operate for unnecessarily long hours to support unnecessarily high loads created by inefficient design, construction and use of the fabric or to support uneconomical equipment which is left on too much."*[15]

To establish realistic figures for actual energy use really does involve accurate arithmetic. This is because every kind of plant and equipment that delivers building services has a fixed energy demand in watts or kilowatts for a given output, an operational efficiency and hours of operation. Count how many there are of each piece of energy-consuming 'kit', multiply each by its hours of use and add them all together.

One of the major reasons for the performance gap as shown earlier in Figure 8 is that the energy models fail to count and add up what have been designated as 'unregulated loads'. This has been confusing the hell out of people for years! If all designs properly accounted for all potential energy demands in a building from the outset, the gap between design and outcome would be significantly reduced.

The tree diagram that lies at the core of TM22: Energy Assessment and Reporting Methodology published by CIBSE[16] graphically illustrates the method. While the tree diagram works downwards to analyse consumption, it can equally be used from the bottom up to build a picture of total realistically expected consumption.

Of course, even if this is done accurately in the first instance, the contractual process must ensure that the concept, procurement and

**FIGURE 11: TM22 tree diagram**

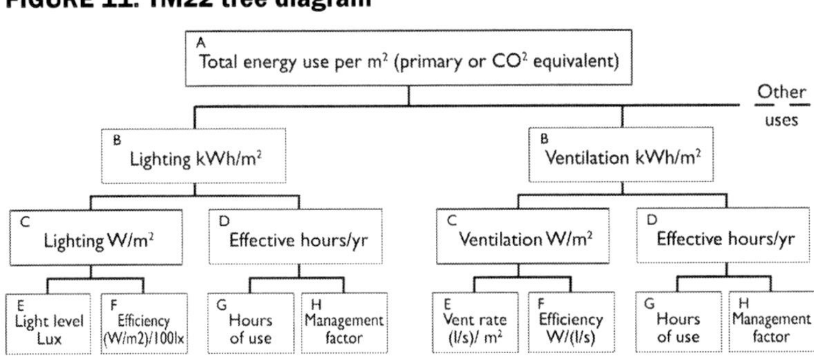


construction teams are required to inform the detailed design team of changes or difficulties on site which may affect the initial estimates. Simply changing 10W light bulbs for 15W ones (because they are cheaper) will lead to a 50% increase in lighting energy demand; the cost consultant, focusing on initial capital cost, will regard this as 'job done', with no appreciation of the significance to future energy costs and carbon emissions. Equally, if the calculation is based on 10 hours of use a day, but they are on for 14 hours a day, then lighting demand will increase by 40%.

Once a building is in use, anyone responsible for specifying and purchasing new kit, should be required to do their sums to establish the impact on the building's energy target, and what action they need to take to ensure that it is not breached.

A similar methodology for estimating demand in the domestic sector has been trialled in the Technology Strategy Board's Building Performance Evaluation (BPE) programme and will shortly be published by CIBSE under the name DomEARM (Domestic Energy and Reporting Methodology).

### 8. Use an integrated process

Develop a process that employs people with the right knowledge and skills, focusses them on outcomes, uses processes that require integrated working by all members of the team, and closes the feedback loop.

### 9. Verify you've got what you asked for – test the building before handover

Checking the quality of the fabric can be done using air leakage tests, thermal imagery and other techniques such as in-situ U-value tests.

Airtightness tests should be conducted at 'first-fix' to identify the location of any significant air leakage paths to ensure they can be more easily remedied. Buildings should then be re-tested at completion, after the services have been installed. All sites should see the installation of site notices such as this...

**FIGURE 12: A new kind of site notice**

**SOURCE:** © Timbertech Homes

All too often, delays in construction mean that there is a rush to complete and hand over the building. Hurried installation of building services can lead to errors, and commissioning can be omitted altogether, with equipment left on its factory default settings – sometimes for the whole life of the equipment.

Thermal imaging can detect leaks in the fabric and heat flows delivered by building services plant and thermal bridging.

To ensure that in-use energy performance can be properly measured and monitored, sub-meters need to be integrated to the building design to allow for the measurement of electrical energy use in different parts of a building and by specific pieces of plant and equipment. Ensuring that metered results are consistent and useful means thoughtful design and careful installation, commissioning, labelling and reconciliation of metering systems.

The term '**soft landings**' refers to a strategy adopted to ensure the transition from construction to occupation is smooth and that operational performance is optimised. Ideally the client should commit to adopting a soft landings strategy in the very early stages so that an appropriate budget can be allocated and contractual arrangements can make the necessary provision. The Soft Landings Framework enables designers and contractors to improve the performance of buildings and generate feedback for project teams over the first two or three years of occupation[17].

10. **Building performance evaluation (BPE)**

Building performance evaluation can be used at various stages of a buildings lifecycle:

- to inform the specification and design of new build, and major retrofit and refurbishment programmes
- to fine-tune new buildings and help with aftercare support
- to improve the design of protoype buildings during the course of phased developments
- to assist with the continuous improvement of existing buildings.

**TABLE 3: The elements of a building performance evaluation process**[18]

| TECHNIQUES | VALUE |
| --- | --- |
| **Pre-visit questionnaire (PVQ)** | Captures basic, background data on a building, including, number of occupants, size, hours of use, and types of mechanical systems |
| **Simple energy assessment** | Data from energy supplier bills is the starting point for understanding building energy use. Digging deeper for greater granularity of energy uses can be problematic unless a robust metering strategy was put in place from the outset. |
| **TM22 Energy Assessment and Reporting Method** | TM22 describes a method for assessing the energy performance of an occupied building and reporting and benchmarking the results. The process can also help improve the investigations and actions required to improve energy efficiency as well as improving building design, management and occupant satisfaction.<br>This is consistent with the methodology for producing a Display Energy Certificate, calculating the sum of delivered energy plus on-site renewables to create a picture of the overall energy demand. |

| TECHNIQUES | VALUE |
|---|---|
| **Semi-structured interview with building managers** | Helps to identify management issues with various building features, such as issues around maintenance, usability and reliability |
| **Building Use Studies (BUS) Occupant Survey** | Helps to identify issues which may be affecting occupants' productivity and effectiveness within a building. This can aid building managers in making adjustments to the operation and controls within the building, thus improving the users' satisfaction with the environmental conditions and use of the building spaces and amenities. |
| **Forensic walk-through of the building** | Helps to identify issues with the building and its operation that can be corrected and improved through management adjustments, controls changes, or-low cost measures such as installing window blinds. |

There are various techniques which are appropriate for gaining useful feedback depending on the stage in the building lifecycle that BPE is undertaken and for which purpose.

## Understanding different types of building

Being able to establish the performance of any building relative to its peers gives a good indication of whether the building is 'good' and is being operated 'well'. This requires the use of benchmarks, which provide representative values for common building types, against which you can compare your building's actual performance. Simple benchmarks of annual energy use per square metre of floor area permit the standard of energy efficiency to be assessed and enable remedial action to be taken. More detailed benchmarks can help pinpoint problem areas within a building.

To be useful, national benchmarks need to be updated regularly with good quality data collected using standard metrics and protocols. Sadly, this has not been the case in the UK; whilst the building benchmarks are held by CIBSE, a professional and trusted independent body, it has not been resourced to help with the gathering of data and making it useful. An increasing number of building owners and occupiers are taking an interest in understanding how their buildings stack up leading to a potential joint undertaking to better maintain UK building benchmark data.

Different building types and users have different energy profiles. The best known benchmarking document is Econ 19 for offices, but that has not been updated since 2003[19]. It divides buildings into 'Typical' and 'Good' practice. Good practice examples are those in which significantly lower energy consumption has been achieved using widely available and well-proven energy-efficient features and management practices. These examples fall within the lower quartile of the data collected. New designs should aim to improve upon the good.

**TABLE 4: Breakdown of energy use by office type**

| | 1 | | 2 | | 3 | | 4 | |
|---|---|---|---|---|---|---|---|---|
| | Good Practice | Typical | Good Practice | Typical | Good Practice | Typical | Good Practice | Typical |
| Heating and hot water - gas or oil | 29 | 151 | 79 | 151 | 97 | 178 | 107 | 201 |
| Cooling | 0 | 0 | 1 | 2 | 14 | 31 | 21 | 41 |
| Fans, pumps, controls | 2 | 6 | 4 | 8 | 30 | 60 | 36 | 67 |
| Humidification (where fitted) | 0 | 0 | 0 | 0 | 8 | 18 | 12 | 23 |
| Lighting | 14 | 23 | 22 | 38 | 27 | 54 | 29 | 60 |
| Office equipment | 12 | 18 | 20 | 27 | 23 | 31 | 23 | 32 |
| Catering, gas | 0 | 0 | 0 | 0 | 0 | 0 | 7 | 9 |
| Catering, electricity | 2 | 3 | 3 | 5 | 5 | 6 | 13 | 15 |
| Other electricity | 3 | 4 | 4 | 5 | 7 | 8 | 13 | 15 |
| Computer room (where appropriate) | 0 | 0 | 0 | 0 | 14 | 18 | 87 | 105 |
| Total gas or oil | 79 | 151 | 79 | 151 | 97 | 178 | 114 | 210 |
| Total electricity | 33 | 54 | 54 | 85 | 128 | 226 | 234 | 358 |

# The Passivhaus energy standard

***"I was working as a physicist. I read that the construction industry had experimented with adding insulation to new buildings and that energy consumption had failed to reduce. This offended me – it was counter to the basic laws of physics. I knew that they must be doing something wrong. So I made it my mission to find out what, and to establish what was needed to do it right."***

**WOLFGANG FEIST, FOUNDER OF PASSIVHAUS INSTITUTE IN GERMANY**

The Passivhaus standard meets the criteria for delivering a building with excellent energy and carbon performance whose in-use performance matches that at the design stage:

| | |
|---|---|
| **An aspirational energy target** | ✔ |
| **An absolute rather than a relative target** | ✔ |
| **Puts delivery of a high quality fabric at the heart of its approach** | ✔ |
| **Accounts for all energy uses and treats the building as an energy system** | ✔ |
| **Creates a common language, and core knowledge and skills for the whole team** | ✔ |
| **Requires a tightly devised specification** | ✔ |
| **Drives energy performance throughout the procurement process** | ✔ |
| **Regularly incorporates monitoring and evaluation of outcomes** | ✔ |

There is a growing vanguard of clients, design teams and builders in the UK and worldwide who are using the Passivhaus standard to given them the confidence that they will acquire and/or deliver buildings with excellent energy and carbon performance.

The standard is being applied in the UK to schools, offices, retail, community buildings, and housing developments. It has proved attractive because it is based on sound building physics and is supported by software (Passivhaus Planning Package or PHPP) that helps design teams to deliver outcomes consistent with design predictions.

Its principal features are:

- space heating/cooling energy use of 15kWh/m$^2$/yr to maintain a temperature of 20°C;

- heat load of 10W/$m^2$;
- airtightness of 0.6 air changes per hour;
- overheating limited to no more than 25 deg C for 10% of total hours;
- a primary energy limit of 120 kWh/$m^2$/yr.

(A primary energy target is the equivalent of setting a carbon limit because it limits the use of electricity, which in the UK and many other countries, is carbon-intensive).

PHPP is based on dynamic thermal simulations validated by data gathered from hundreds of low-energy buildings. Its level of sophistication enables it to take into account a wide range of variable characteristics which affect heat loss and subsequent energy use (including lighting, equipment and appliances), allowing for a superior fit between predicted energy use and real-world performance. It is suitable for both residential and many non-residential applications and is appropriate for use as a design tool which significantly aids those on the design team to understand the effect of even minor changes on likely out-turn energy performance. The model constitutes 35 Excel spreadsheets, which require the provision of much more detailed data than SAP or SBEM, but with a resulting big improvement in accuracy.

The Passivhaus Institute (PHI) certifies designers and contractors who have undertaken their training course and passed a rigorous assessment to demonstrate their understanding of the principles and practices required to deliver a very low energy building. It certifies products that meet its strict in-use performance criteria. Many of its buildings are subsequently evaluated to establish whether the design intent has been delivered in practice and it makes the results of its evaluations public.

The Passivhaus methodology uses integrated design and construction processes, involving contractors at an early stage. It makes it possible to create a building with much better energy and environmental performance at virtually the same cost as a conventional building.

Interserve built its new head office building to the Passivhaus standard

Occupant Feedback:

> ***'We were surprised to realise that since moving to our Passivhaus office our staff sickness days have fallen 13% compared with our previous office. Assuming that this is repeated in schools and other commercial buildings, it provides another demonstrable benefit of Passivhaus.'***
>
> **ANDREW HOWARD – REGIONAL DIRECTOR**
>
> ***The tranquillity, light quality, and constant temperatures are a far cry from our old office where we froze in winter and roasted in summer in a dark and draughty old building.'***
>
> **DANIEL MURDOCK – BUSINESS DEVELOPMENT COORDINATOR**
>
> ***The only way which we could afford to move to new offices was the Passivhaus route, the equivalent BREEAM may have delivered similar quality but would have cost a lot more to run. We have managed to move from a tired old office with massive energy bills to a wonderful new office where everyone enjoys the pleasant environment with negligible energy bills at no additional cost – No Contest!'***
>
> **GORDON KEW – REGIONAL BUILDING DIRECTOR**

## Establishing your own plan of action

1. **Identify where you sit on the buildings cycle diagram**

What you can do to apply what you have learned depends on where you can make a difference – either by making changes yourself or by influencing change in others or in your organisation's strategic objectives and processes in relation to buildings.

2. **Whatever role you play:**
    - Understand what information you need and can access
    - Maximize energy savings and minimize frustration by using best practices
    - Determine what energy uses to measure and how best to measure them to characterize building performance
    - Calculate expected energy and cost savings and resulting values
    - Compare expected performance with measured performance to diagnose problems and identify additional savings opportunities.

3. **You are all building occupants, so you can probably all start with looking more carefully at your own buildings – never mind how small they are, or how energy-intensive ...**
    - Do you have the information in $kWh/m^2/yr$ for space and water heating and electricity?
    - Is this good or bad?
    - Could it be better?
    - If you've got sub-meters, reconciling them would be a good place to start

***How do you compare with other buildings of the same type?***

- Check out building benchmarks for your type of building

**TABLE 5: Sample of benchmarks from TM46 Energy Benchmarks**[10]

| Building category | Electricity typical benchmark (annual kWh/m²) | Fossil-thermal typical benchmark (annual kWh/m²) |
|---|---|---|
| General office | 95 | 120 |
| General retail | 165 | 0 |
| Large food store (supermarket) | 400 | 105 |
| Hotel | 105 | 330 |
| Schools and seasonal public buildings | 40 | 150 |
| University campus | 80 | 240 |
| Hospital (clinical and research) | 90 | 420 |
| Storage facility – storage warehouse or depot | 35 | 160 |

Floor area is generally based upon Gross Floor Area of building

***How much energy does you building use when you are not using it?***

- There are easy tests that quickly signpost unnecessary energy use or energy waste such as monitoring daily energy use and spotting that energy consumption does not fall at weekends, for example.

> ***"I went in at the weekend when I knew I could turn off absolutely everything without interfering with the staff's activities.That's when I established that the school next door was connected to our meter and had been for the past seven years."***
>
> **SCHOOL CARETAKER!**

- Turn things on and off and check the meter, plug-in ones for plug loads.
- Start counting and adding up your bits of kit and their energy demands.
- What kind of building you have got? Have you been dealt a good hand of cards or a bad one?
- How does it compare with the characteristics of a low energy building in Section 3?

***Could you think laterally about solutions to problems that actually reduce energy demand?***

> ***"I was asked to install air-conditioning to reduce overheating in the computer room. I was really struggling to find something that would work. Then it occurred to me – it was the computers creating all the heat! I got permission to replace the computers with energy-efficient ones to eliminate the overheating problem and reduce our energy costs all round."***
>
> **A DIFFERENT SCHOOL CARETAKER!**

***Is your building one that is comfortable and productive that its occupants like?***

- Diagnose your building to establish where it can be made better.
- Why not evaluate the performance of your building to really understand what you have got?
- Why not test the fabric using thermal imaging and an airtightness test?

***Who is responsible for managing your building energy?***

- Do they have the information and the authority to manage energy use down?
- Do you have a clear absolute target for your energy use?
- Are all the occupants contributing to meeting them?

If you are about to embark on specifying, procuring or selecting new premises or undertaking a major refurbishment, draw insights from studying your existing premises and use these to help get the right type and performance of building for your future needs.

Finally, and as importantly, how could you contribute to the knowledge base by making your energy data public to contribute to better benchmarks? New tools and techniques, databases and so on, will be appearing soon so keep an eye on developments.

PART 5

# Why Bother With All This? Is Anyone Else?

**"There is growing evidence that sustainable buildings are more attractive to both occupiers and investors, as they are more efficient to operate and tend to age better."**

ADRIAN PENFOLD, HEAD OF PLANNING AND CORPORATE RESPONSIBILITY AT BRITISH LAND[20]

## The business case for buildings with better energy and carbon performance

**PART ONE ARGUED THE CASE** for buildings with better energy and carbon performance on the grounds of improving productivity and business continuity for their occupants, and an increase in reputational capital – for those who own them, and for those who contribute to their design, construction and operation. There are also sound financial reasons for having better buildings. As British Land argues above, sustainable buildings are likely to hold or increase their value by being more attractive to investors and occupiers.

Lower running costs can also play their part. Saving money on energy is not normally at the forefront of business managers' minds when energy costs usually represent a small fraction of total operating costs. On the other hand, when Energy UK, the industry trade association, predicts price rises of 50% between now and 2020[21], every financial director should be concerned at a controllable cost suddenly looking out-of-hand. And be wondering what other kinds of turbulence might accompany such massive price rises – the brown-outs, power cuts – or a three-day week!

But there are also many opportunities to pick low-hanging fruit. The Better Buildings Partnership and Jones Lang LaSalle undertook analysis of the energy use of a range of commercial office buildings ranked by their EPC rating – demonstrating unequivocally that an EPC rating bears little correlation to actual energy use. Indeed, some of the better-rated buildings by EPC had significantly higher energy use than the worst-rated buildings. The scope for improved building operation and cost savings is self-evident, even in air-conditioned buildings.

**FIGURE 13: Energy intensity of buildings by EPC rating**

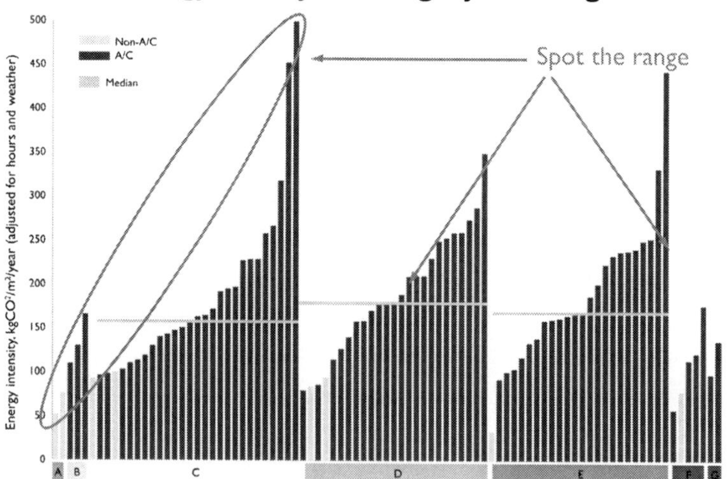


**SOURCE:** Courtesy Better Buildings Partnership, from Jones Lang Lasalle, ***A Tale of Two Buildings***

Regulatory constraints are also closing in. Companies are increasingly required to report on their energy use and carbon emissions. These include:

- CRC Energy Efficiency Scheme
- Climate Change Agreements
- the EU Emissions Trading System
- mandatory greenhouse gas reporting

The Energy Saving Opportunity Scheme (ESOS)[22] requires each government in the European Union to introduce a programme of regular energy audits in every enterprise employing more than 250 people or turning over above €50m a year to cover every aspect of any business. The first set of audits must be completed by December 2015 and must be repeated every four years. Clear information must be provided, identifying and quantifying cost-effective energy saving opportunities based on life-cycle measurements rather than simple payback periods. (The Energy Efficiency Deployment Office at DECC intends to ensure that the reporting mechanisms will dovetail with those already in force, to avoid unnecessary duplication for participating enterprises).

And from 2018, it will no longer be possible to sell or rent buildings with an EPC rating of F or G. Who wants to be left holding the dregs of the property sector?

## What are others doing? Acquisition, specification and preliminary design

Client Bath and North East Somerset Council made it clear from the outset that it was looking for its new buildings in the Keynsham regeneration area to offer higher-than-average energy-efficiency and sustainability credentials. It sought out a project team that would deliver a building offering the very highest energy-efficiency credentials on a sound budget. The contract was won by Willmott Dixon, who explained: "*We were targeted to deliver an Energy Performance Certificate/Display Energy Certificate operational rating of A. The new building will be 90 per cent more efficient than the old one. This is a clever design and to deliver that you need a client which understands exactly what is required.*"[23]

Wolverhampton City Council has procured three Passivhaus schools and a university in the Midlands is in the process of delivering a medical research centre to the standard. For others working to the standard, see **www.passivhaustrust.org.**

## What are others doing? Owner-managers

*Refurbishing the worst properties:* The Climate Change Property Fund specialises in investing in buildings whose DEC rating can be significantly improved. Its investment at 5 St Philips Place, Birmingham, contributed towards a 55% saving in tenant energy bills over four years. CCPF reduced the DEC from a G to a C through the installation of LED lights, and modification of air conditioning and metering. The display of half-hourly readings for electricity, gas, water and carbon on a television monitor in the lobby helped focus the behaviour of occupiers. CCPF set up a green lease with the tenant, requiring them to invest in smart metering, controls, and light reconfiguration. It demonstrates payback of only three years by quantifying the benefits using the CIBSE TM22 energy assessment tool.

> ***"The benefits for us a re a happy tenant, who in turn benefits from the majority of energy savings. They will be encouraged to***

***stay with us for a long time and renew the lease."***

**TIM MOCKETT, JOINT MANAGING DIRECTOR**[24]

Jones Lang LaSalle undertook extensive analysis of the energy and investment performance over the 15 months up to July 2011 of another Climate Change Property Fund building, 77 Gracechurch Street. They compared it with two directly comparable, but less sustainable, properties in the City of London. JLL concluded that active energy management alongside additional factors such as tenant goodwill, lower letting risk, lower exposure to carbon tax, stronger defensive position at rent review, created a 'green alpha' of some 15-20% of total performance.

## What are others doing? Sales and marketing

How can you get a good building when you are looking to rent or buy? What should you be asking?

The Better Buildings Partnership (www.betterbuildingspartnership.co.uk) is a collaboration of the UK's leading commercial property owners who are working together to improve the sustainability of existing commercial building stock. It has taken on major challenges facing the property market and attempts to develop a solution or toolkit which can help the market overcome them.

It has published a number of free tools, including:

- A Green Lease Toolkit
- Better Metering Toolkit
- Sustainability Benchmarking Toolkit

- Managing Agents Sustainability Toolkit
- Transactional Agents Sustainability Toolkit
- Green Building Management Toolkit
- Landlord Energy Rating

Better Buildings Partnership: **http://www.betterbuildingspartnership.co.uk**

## What are others doing?
## Detailed design and construction

Many *architects* are undertaking training to get a better understanding of how to deliver low energy buildings, generally in terms of working to the Passivhaus standard. This requires working to absolute outcomes and encourages monitoring actual energy use.

CarbonBuzz is a free on-line resource which allows designers to compare predicted energy performance with actual. The database can be browsed without the need for registering or logging in; once registered, project templates can be created, allowing users to add multiple records, which can be drawn from any project stage from design inception to operation. Records can be created with a minimum of total gas and electricity for the year or users can drill down to finer granularity if they choose. CarbonBuzz allows users to analyse and compare data entered for their buildings at each project stage through energy and carbon graphs known as Energy Bars. Users can track, review and compare energy records and contributing factors over time and generate reports on their data for external use. All data can remain anonymous. However, users can choose to share their CarbonBuzz projects, records or portfolios with specified users and organisations to facilitate design and operation.

Very few architects yet promote themselves as being able to deliver low energy buildings. Those that do tend to use their ability to design to the Passivhaus energy standard, offering design consultancy, energy analysis, project enabling and training for clients, design teams and constructors.

## What are others doing? Major contractors

Originally six, now eleven major contractors, are partners in Supply Chain Sustainability School, with significant funding from CITB. The School represents a common approach to addressing all aspects of sustainability within their supply chains, including energy, and provides self-assessment and learning using a range of free tools, e-learning modules and sustainability training. **www.supplychainschool.co.uk**

## What are others doing? Occupation, management and maintenance

*Going public voluntarily:* Some government departments – and gradually owners of other types of building – have been at the forefront of reducing their energy use and carbon emissions and going public with the results – **www.carbonculture.net**. They not only show their energy use in real time, but are able to demonstrate the effectiveness of the actions that they have taken.

> ***"The first time I got there and the door was closed, I thought 'Blimey, DECC's been closed down!' But of course, it was just part of its new energy-saving strategy."***

Display Energy Certificates have been useful in making energy use and carbon emissions visible. The government's recent decision to comply only with the minimum requirements of the Energy Performance of Buildings Directive means that DECs will only need to be up-to-dated every ten years and smaller buildings will be required to display only an Energy Performance Certificate. Failure to implement the option of extending DECs to the private sector is seen as a missed opportunity and there have been moves on the part of some national organisations to encourage the use of voluntary DECs so that we continue to build the knowledge base. This has sometimes been dubbed the Operational Energy Rating or OER. We prefer to call it the 'OoER'! It has the advantage that it can convey both "Wow, that's fantastic, we're doing a great job!" or "Oh dear, we seem to have dropped the ball on this one."

**The fact is that improving the energy and carbon performance of buildings is an issue that is here to stay. We are on the cusp of radical change to which the organisations listed above are testament. It's no longer a question of "Are you going to be a part of this?" but "Why aren't you doing this now?"**

# Notes and References

1. The requirement for larger commercial properties to display a DEC was repealed by the Energy Act 2011.
2. Urban Land Institute and Cambridge University Centre for Sustainable Development, 2013, *Green premium or grey discount?: The value of green workplaces for commercial building occupiers in the UK.*
3. **http://chp.decc.gov.uk/cms/centralised-electricity-generation**
4. Energy Saving Trust, 2010. *Getting warmer: a field trial of heat pumps report.*
5. Energy Saving Trust, 2013. *The heat is on: heat pump field trials: phase 2.*
6. PROBE was a research project conducted by Building Services Journal and managed by HG Consulting Engineers, Building Use Studies, William Bordass Associates and Building Services Journal. The PROBE research was co-funded under the Partners in Technology collaborative research programme run by the Department of the Environment, Transport and the Regions.
7. The PROBE Team, 1999 *Probe Review Final Report 4, Strategic conclusions* (p11).
8. CarbonBuzz (**www.carbonbuzz.org**).
9. EMBED is still in development at May 2014 at **www.getembed.com.**
10. CIBSE, *TM46 Energy benchmarks.*
11. *Building Use Surveys* **http://www.busmethodology.org.uk**
12. *The Guardian*, 7 September 2013.
13. Building Services Journal, April 1998, Elizabeth Fry PROBE study.

14. Malcolm Bell, 2013. Achieving near zero carbon housing: The role of performance measurement in production processes. *Workshop on High Performance Buildings; Design and Evaluation Methodologies, 24 – 26 June 2013, Belgian Building Research Institute, Brussels.*

15. The PROBE Team, 1999 *Probe Review Final Report 4, Strategic conclusions P13.*

16. CIBSE, TM22: Energy Assessment and Reporting Methodology.

17. Soft Landings, **https://www.bsria.co.uk/services/design/soft-landings**

18. Institute for Sustainability, 2013, *Guide to Building Performance Evaluation.*

19. Energy Consumption Guide 19: Energy efficiency in offices, 2003 edition.

20. British Land, 2011. *Sustainability Brief for Developments.*

21. *Guardian*, 12 November 2013.

22. Energy Saving Opportunities Scheme – Article 8 of the Energy Efficiency Directive, **http://ec.europa.eu/energy/efficiency/eed/eed_en.htm**

23. *Construction News*, 20th November 2013.

24. Life after DECs, *CIBSE Journal*, February 2013.